Edexcel Award in

Algebra

Su Nicholson

ALWAYS LEARNING

PEARSON

Published by Pearson Education Limited, Edinburgh Gate, Harlow, Essex, CM20 2JE.

www.pearsonschoolsandfecolleges.co.uk

Edited by Project One Publishing Solutions, Scotland
Typeset and illustrated by Tech-Set Ltd, Gateshead

First published 2013

17 16 15 14 13
10 9 8 7 6 5 4 3 2 1

British Library Cataloguing in Publication Data
A catalogue record for this book is available from the British Library

ISBN 978 1 446 90322 3

Printed in Slovakia by Neografia

Acknowledgements
Every effort has been made to contact copyright holders of material reproduced in this book. Any omissions will be rectified in subsequent printings if notice is given to the publishers.

Notices
The GCSE links provide references to course books as follows:
AF Edexcel GCSE Mathematics A Foundation Student Book
AH Edexcel GCSE Mathematics A Higher Student Book
BF Edexcel GCSE Mathematics B Foundation Student Book
BH Edexcel GCSE Mathematics B Higher Student Book
16+ Edexcel GCSE Mathematics 16+ Student Book

Contents

Name .... Class ....

Self-assessment chart

	Needs more practice	Almost there	I'm proficient!	Notes
Chapter 1 Roles of symbols				
1.1 Using letters to represent numbers	☐	☐	☐	
1.2 Equations, formulae and expressions	☐	☐	☐	
1.3 Representing situations in real life	☐	☐	☐	
Chapter 2 Algebraic manipulation				
2.1 Collecting like terms	☐	☐	☐	
2.2 Multiplication with brackets	☐	☐	☐	
2.3 Factorising	☐	☐	☐	
2.4 Laws of indices	☐	☐	☐	
Chapter 3 Formulae				
3.1 Using word formulae	☐	☐	☐	
3.2 Using algebraic formulae	☐	☐	☐	
3.3 Changing the subject of a formula	☐	☐	☐	
Chapter 4 Linear equations				
4.1 Solving equations with one operation	☐	☐	☐	
4.2 Solving equations with two operations	☐	☐	☐	
4.3 Solving equations with brackets	☐	☐	☐	
4.4 Solving equations where the unknown appears on both sides	☐	☐	☐	
4.5 Solving equations with negative coefficients	☐	☐	☐	
Chapter 5 Linear inequalities				
5.1 Introducing inequalities	☐	☐	☐	
5.2 Representing inequalities on a number line	☐	☐	☐	
5.3 Solving linear inequalities	☐	☐	☐	
5.4 Finding integer solutions to inequalities	☐	☐	☐	
Chapter 6 Number sequences				
6.1 Term-to-term and position-to-term rules	☐	☐	☐	
6.2 The nth term of an arithmetic sequence	☐	☐	☐	
Chapter 7 Gradients of straight line graphs				
7.1 Finding the gradient	☐	☐	☐	
7.2 Interpreting the gradient	☐	☐	☐	
Chapter 8 Straight line graphs				
8.1 Horizontal and vertical lines	☐	☐	☐	
8.2 The equation $y = mx + c$	☐	☐	☐	
8.3 Plotting and drawing graphs	☐	☐	☐	
8.4 Finding the equation of a straight line graph	☐	☐	☐	
Chapter 9 Graphs for real-life situations				
9.1 Straight line graphs	☐	☐	☐	
9.2 Different graph shapes	☐	☐	☐	
Chapter 10 Graph sketching				
10.1 Quadratic graphs	☐	☐	☐	
10.2 Graphs of the form $y = ax^2$	☐	☐	☐	
10.3 Graphs of the form $y = ax^2 + b$	☐	☐	☐	
10.4 Graphs of the form $y = (x + b)^2$	☐	☐	☐	
Chapter 11 Simple quadratic functions				
11.1 Plot graphs of quadratic functions of the form $y = ax^2 + b$	☐	☐	☐	
11.2 Plot graphs of quadratic functions of the form $y = ax^2 + bx + c$	☐	☐	☐	
11.3 Using quadratic graphs to solve equations	☐	☐	☐	
Chapter 12 Distance–time and speed–time graphs				
12.1 Speed	☐	☐	☐	
12.2 Distance–time graphs	☐	☐	☐	
12.3 Speed–time graphs	☐	☐	☐	

Needs more practice ☐ Almost there ☐ I'm proficient! ☐

1.1 Using letters to represent numbers

GCSE LINKS
AF: 4.1 Using letters to represent numbers; **BF:** Unit 2 7.2 Using letters to represent numbers

By the end of this section you will know how to:

* Use letters instead of numbers

Key points

* **Algebra** is the part of mathematics where letters and symbols replace numbers in order to represent relationships that can vary (over time).
* Letters can be used instead of numbers to represent situations.
* In algebra **simplify** means to write something in a shorter form.

Remember this

* a is the same as $1a$: by convention we don't write the 1
* $2 \times a = 2a$: we don't write the $\times$ symbol
* $a + a = 2a$, not $a2$: always write the number in front of the letter.

Guided

1 Simplify

a $a + a + a + a$

$a + a + a + a = 4$ ....

b $x + x + x$

$x + x + x =$ x

c $5 \times c$

$5 \times c = 5$ ....

d $b \times 3$

$b \times 3 =$ ....

2 Simplify

a $3a + 4a$

$3a + 4a = 7a$

+3 +4
−1 0 1 2 3 4 5 6 7

b $5b - b$

$5b - b = 5b - 1b = 4$ ....

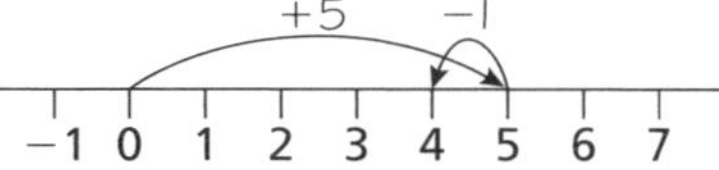

c $4c + 2c - 7c$

$4c + 2c - 7c =$ ....

−1 0 1 2 3 4 5 6 7

You should know

Use a number line to add or subtract the numbers in front of the letter.

Practice

3 Simplify

−5 −4 −3 −2 −1 0 1 2 3 4 5 6 7 8 9 10 11 12 13 14 15

Hint

Use the number line to help.

a $k + k + k + k + k + k + k$

b $8 \times y$

c $4x + 7x$

d $9y - 11y$

e $12k - 4k - k$

f $3m + m - 9m + 2m$

4 Simplify

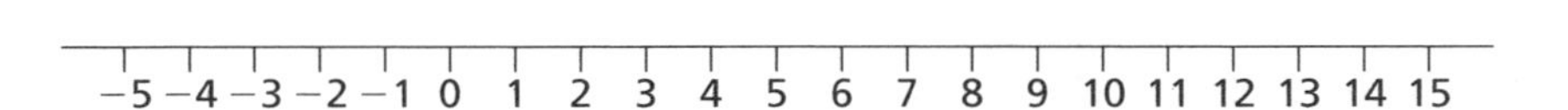

a $2a + 3a + 7a$

b $4b + 2b - 5b$

c $8c - 2c - 3c$

d $6d - 2d - 4d - d$

Needs more practice ☐ Almost there ☐ I'm proficient! ☐

1.2 Equations, formulae and expressions

By the end of this section you will know how to:

* Understand the difference between an equation, a formula and an expression

GCSE LINKS
AH: 13.6 Distinguishing between 'equation', 'formula', 'identity' and 'expression';
BH: Unit 2 10.1 Distinguishing between 'equation', 'formula', 'identity' and 'expression'

Expressions

Key points

* An **algebraic expression** is a collection of **terms**. For example, $2x + 3y - 7$ is an algebraic expression and $2x$, $3y$ and -7 are all terms in this expression.
* In the above expression, x and y are called **variables** as their values can change and -7 is a called a **constant**.

You should know
A variable can be represented by an upper or lower case letter, e.g. A or a.

Guided

1 Which of the following are algebraic expressions?

a $4p - 3s + 5$ **b** $3x = 7y + 9$ **c** $3 \times a \times b$

a and .... are algebraic expressions.

Remember this
There is no 'equals' sign (=) in an algebraic expression.

Formulae

Key points

* A **formula** (plural **formulae**) is a mathematical relationship or rule expressed in words or symbols. For example, the formula for the area, A, of a circle with radius, r, is $A = \pi r^2$.

Guided

2 Here are some formulae. Match each description to the correct formula.

a area, A, of a triangle $P = 2(l + w)$

b perimeter of a rectangle $V = lwh$

c volume of a cuboid $A = \frac{1}{2}bh$

Remember this
A formula has an 'equals' sign (=).

Equations

Key points

* An **equation** is a mathematical statement that shows the equality of two expressions.
* An equation can be solved to find the value of the variable. For example, $x + 2 = 7$ is true for $x = 5$ as $5 + 2 = 7$.
* A **formula** and an **equation** both contain an 'equals' sign.

Guided

3 State whether each of the following is an equation, a formula or an expression.

a $3x - 6 = 12$ $3x - 6 = 12$ is an e..........................

b $3y - 7x + 5$ $3y - 7x + 5$ is an e..........................

c $C = \pi d$ $C = \pi d$ is ....

Practice

4 State whether each of the following is an equation, a formula or an expression.

a $x^2 - 1 = 8$

b $v = u + at$

c $4x + 1 = 2x - 5$

d $10p - 7q + 36$

5 State whether each of the following is an equation, a formula or an expression.

a $4P - 8 = 20$

b $2s - 5t + 8$

c $E = mc^2$

d $C = 2\pi r$

Needs more practice ☐ Almost there ☐ I'm proficient! 

1.3 Representing situations in real life

GCSE LINKS
AH: 2.1 Collecting like terms; **BH:** Unit 2 7.1 Collecting like terms

By the end of this section you will know how to:

* Write an expression to represent a situation in 'real life'

Key points

* You can use letters to represent values that may vary in a real situation.
* It can be helpful to use a letter that links to the real-life item it represents, for example, b batteries, s sweets, d DVDs, and so on.

Guided

1 Paul has b batteries. He buys 4 more batteries.
How many batteries has he got altogether? $b +$ ....

2 Eva buys a bag of s sweets. She eats 2 of them.
How many sweets does she have left? $s -$ ....

Practice

3 Bilal has d DVDs. He buys 6 more. How many DVDs does Bilal have now?

4 Jia buys a apples. She gives 3 of them to her friends. How many apples does Jia have left?

5 Claire is y years old. Tim is 3 years younger. How old is Tim?

6 Bill and Harry go shopping. Bill buys x pairs of socks and Harry buys y pairs of socks.
How many pairs of socks do they buy altogether?

7 Fred is m years old. Sally is 3 times as old as Fred.
Write down an expression, in terms of m, for Sally's age.

Guided

8 Helen sells chocolates. She sells chocolates in boxes of 6 or in boxes of 12.
One day she sells p boxes of 6 chocolates and q boxes of 12 chocolates.
How many chocolates does she sell altogether?

Number of chocolates in p boxes of 6 chocolates $= 6 \times p =$ ....

Number of chocolates in q boxes of 12 chocolates $= 12 \times q =$ ....

Total number of chocolates = ... + ...

Remember this

Different letters cannot be combined or simplified. For example, $6p + 12q$ cannot be simplified.

Practice

9 Batteries are sold in packs of 4 and packs of 8.
Vernon buys x packs of 4 batteries and y packs of 8 batteries.
How many batteries does he buy altogether?

10 Cupcakes are sold in boxes of 4 and boxes of 9.
One day Alisha sells e boxes each containing 4 cupcakes and f boxes each containing 9 cupcakes.
How many cupcakes does she sell altogether?

11 Mo sells bunches of roses.
He sells a small bunch of roses for £5 and a large bunch of roses for £8.
One day Mo sells g small bunches of roses and h large bunches of roses.
How much money, in pounds, does he receive?

Remember this

In Q11 you are asked for the answer in pounds, but you do not need to write the £ symbol in your expression. In general you should keep units out of expressions, formulae and equations.

Step into GCSE

12 A café sells cups of coffee for 90p and cups of tea for 75p.
On one day the café sells c cups of coffee and t cups of tea.
Write down an expression for the amount of money, in pounds, the café received for these sales.

Don't forget!

* Match each of the following with the correct description:

$P = 2l + 2w$	$8 + 4a - 5b$	$3x - 7 = 2$
equation	formula	expression

* Complete the following:

8, $4a$ and $-5b$ are called .... of the .... $8 + 4a - 5b$.

* Mia is y years old. Sasha is 2 years older than Mia. Ella is 1 year younger than Mia. So:

Sasha is .... years old Ella is .... years old

* Simplify the following:

$4x + 5x - 6x =$ .... $3y - y - 6y =$ ....

Exam-style questions

1 Simplify $5a - 2a + 6a$

2 Simplify $2y - y - 7y$

3 Three of the following are formulae. Tick (✓) them.

$y^2 + 2 = 27$ ☐ $M = DV$ ☐ $10 - a = 6$ ☐

$a^2 + b^2 = c^2$ ☐ $3x - 4 + 7y$ ☐ $A = lw$ ☐

4 Three of the following are equations. Tick (✓) them.

$5 - x^2 + 2y$ ☐ $V = LWH$ ☐ $12k^2 + 7 = 33$ ☐

$14 = 5x - 1$ ☐ $y = mx + c$ ☐ $4x - 16 = 0$ ☐

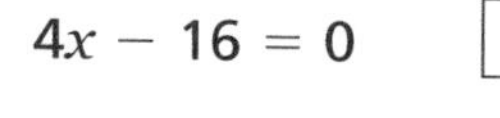

5 Aki buys b bottles of water. Dai buys w bottles of water.
How many bottles of water do Aki and Dai buy altogether?

6 Sam is c cm tall. Tim is 5 cm taller. How tall is Tim in cm?

7 Kelly weighs p pounds. Pat is 3 pounds lighter. How much does Pat weigh in pounds?

8 Emma weighs k kilograms. Her son, Ben, weighs $\frac{1}{3}$ of Emma's weight.
How much, in kilograms, does Ben weigh?

9 Pens are sold in packets of 3 and in packets of 10. Kevin buys p packets containing 3 pens and q packets containing 10 pens. How many pens does Kevin buy altogether?

10 A shop sells DVDs for £6 and CDs for £4. Brodie buys d DVDs and c CDs.
How much money, in pounds, does Brodie pay for these?

11 Khalid sells bottles of water for 80p. He sells cans of cola for 95p.
On one day Khalid sells w bottles of water and c cans of cola. Write down an expression, in pounds, for the amount of money Khalid receives for these sales.

Needs more practice ☐ Almost there ☐ I'm proficient! ☐

2.1 Collecting like terms

GCSE LINKS
AH: 2.1 Collecting like terms;
BH: Unit 2 7.1 Collecting like terms;
16+: 6.1 Collect like terms

By the end of this section you will know how to:

* Manipulate algebraic expressions by collecting like terms

Key points

* Terms that use the same variable or letter or combination of letters are called **like terms**. For example, x and $3x$, $5y^2$ and $-7y^2$, $4xy$ and $-2xy$.
* You can add and subtract like terms to **simplify** expressions.
* Different letter symbols cannot be added or subtracted. For example, $2x + 5y$ cannot be simplified further.
* The **sign** of a term in an expression is always written before the term. For example, in the expression $6 + 5x - 2y$, the '+' sign means add $5x$ and the '−' sign means subtract $2y$.

You should know
A **variable** is a value that can change.

Guided

1 Simplify

a $2x - 8 + 7x - 6$

$= 2x + 7x - 8 - 6$

$= 9x -$ ....

b $3a + 4b - a - 7b$

$= 3a - a + 4b - 7b$

$= 2a$ ....

c $9 - 8c + 5d + 2 - 4c + 3d$

$= 9 + 2$ ....

$=$ ....

Hint
It is usual to avoid writing the final answer starting with a '−' sign if possible. For example, $-6x + 5$ would be rewritten as $5 - 6x$

Remember this
You can re-write algebraic expressions so that like terms are next to each other.

Practice

2 Simplify

a $3x - 7y + 2 - 5x + 10y - 9$

b $7p - 8r + 9 - 2p + 11r - 4$

c $a - 6b + 9 - 4a + 8b - 11$

d $3s + 5t - 7 - 4s - t - 4$

Guided

3 Simplify

Remember this
$x^2 = x \times x$: the '2' tells you how many xs are multiplied together, so $x^3 = x \times x \times x$.

You should know
$xy = x \times y$: if there is no sign between the letters this means they are multiplied together.

a $3x + x^2 - 2x - 4x^2$

$= 3x - 2x + x^2 - 4x^2$

$=$ .... $-$ .... x^2

Hint
An answer such as $3x^2 + 2x$ cannot be simplified.

b $3xy - 10 - xy + 7$

$= 3xy - xy - 10 + 7$

$=$ ....

c $3x^2 + 5y^2 - 9xy - 2xy - 7y^2 + 6x^2$

$= 3x^2$ ....

$=$ ....

d $7y^2 + 2x^2 + xy - 4y^2 - 3xy - 5x^2$

$=$ ....

$=$ ....

Practice

4 Simplify

a $5k^2 + 7k - 3 + 2k^2 - 6k + 2$

b $8vw + 9w - 3vw + 5 - 2w + 10$

c $2g^2 - 3h^2 + 1 - 5g^2 + 9h^2 - 4$

d $5a^2 - 4ab + 3b^2 + 8 - 2a^2 - 7ab - 4b^2$

Step into GCSE

5 Simplify

a $6e^3 + 7e^2 - e - 2e^3 - 3e^2 + 9e$

b $5c^2 - 2c - 1 + 7c^2 - 6c + 9$

6 The diagram shows a triangle.
Write down an expression, in terms of x and y, for the perimeter of the triangle.
Simplify your answer.

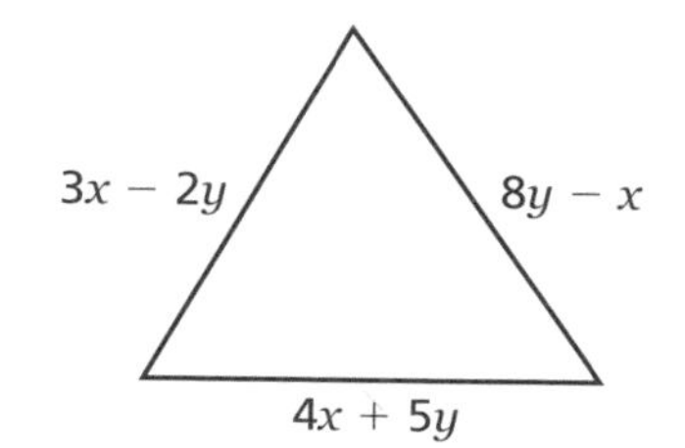

Hint
Add together the lengths of the sides and then collect like terms.

Needs more practice ☐ Almost there ☐ I'm proficient! ☐

2.2 Multiplication with brackets

By the end of this section you will know how to:

* Multiply a single term over a bracket

GCSE LINKS
AH: 9.1 Expanding brackets; **BH:** Unit 2 8.1 Expanding brackets; **16+:** 6.3 Expand single brackets

Key points

* Multiplying a term over a bracket is called **expanding** the bracket.
* When there is a number or letter next to a bracket, this means there is a hidden multiplication sign. For example, $2(x - 4) = 2 \times (x - 4)$.
* Everything inside the bracket must be multiplied by whatever is outside the bracket.

You should know
When multiplying (or dividing) positive and negative numbers, if the signs are the same the answer is '+'; if the signs are different the answer is '−'.

Guided

1 Expand

a $5(y + 8)$
$= 5 \times y + 5 \times 8$
$= 5y +$ ....

b $3(2x - 1)$
$= 3 \times 2x$ .... $3 \times$ ....
$=$ ....

c $4(6a - 3b)$
$= 4 \times$ ....
$=$ ....

d $-2(5pq + 4q^2)$
$= -2 \times 5pq -$ ....
$=$ ....

e $-3(4w^2 - 7v^2)$
$=$ ....
$=$ ....
$=$ ....

f $-(3xy - 2y^2)$
$=$ ....
$=$ ....
$=$ ....

Hint
$-(3x - 2y)$ is the same as $-1(3x - 2y)$

2 Expand and simplify

> **Remember this**
> First expand the brackets, then collect like terms.

a $7(3x + 5) + 6(2x - 8)$

$= 21x + 35 + \ldots x - \ldots$

$= \ldots x - \ldots$

b $8(5p - 2) - 3(4p + 9)$

$= 40p - \ldots - \ldots p \ldots$

$= \ldots p \ldots$

c $9(3s + 1) - 5(6s - 10)$

$= \ldots$

$= \ldots$

$= \ldots$

d $2(4x - 3) - (3x + 5)$

$= \ldots$

$= \ldots$

Practice

3 Expand

a $-2(3b^2 - 7)$

b $5(3pq - p^2)$

c $-4(2x^2 - 3y^2)$

d $3(7a^2 - 5b^2)$

4 Expand and simplify

a $4(3k - 2) + 5(2k + 7)$

b $3(7y + 1) - 6(4y + 5)$

c $2(9 - 4x) - 7(5 - 3x)$

d $6(4a - 3) - (2a + 7)$

Guided

5 Expand

> **Remember this**
> $x \times x^2 = x \times x \times x = x^3$

a $2a(3a - 7)$

$= 2a \times 3a + 2a \times -7$

$= 6a^2 \ldots$

b $3r(4r^2 - 8)$

$= 3r \times 4r^2 \ldots 3r \times \ldots$

$= 12 \ldots$

c $-5s(2s^2 - 9s + 10)$

$= \ldots$

$= \ldots$

d $-3x(2x^2 - 7x + 5)$

$= \ldots$

$= \ldots$

Practice

6 Expand

a $3x(4x + 8)$

b $4k(5k^2 - 12)$

c $-2h(6h^2 + 11h - 5)$

d $-3s(4s^2 - 7s + 2)$

7 Expand and simplify

a $3(y^2 - 8) - 4(y^2 - 5)$

b $2x(x + 5) + 3x(x - 7)$

c $4p(2p - 1) - 3p(5p - 2)$

d $3b(4b - 3) - b(6b - 9)$

GCSE

8 Expand $\frac{1}{2}(2y - 8)$

9 Expand and simplify

a $13 - 2(m + 7)$

b $5p(p^2 + 6p) - 9p(2p - 3)$

10 The diagram shows a rectangle.
Write down an expression, in terms of x, for the area of the rectangle.
Show that the area of the rectangle can be written as $21x^2 - 35x$.

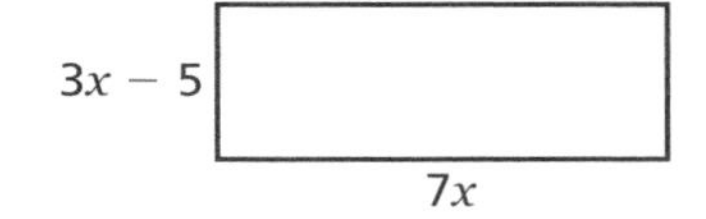

Hint
The area of a rectangle = length × width.

Needs more practice ☐ Almost there ☐ I'm proficient! ☐

2.3 Factorising

By the end of this section you will know how to:

* Factorise by taking out common factors

GCSE LINKS
AH: 9.2 Factorising by taking out common factors; **BH:** Unit 2 8.2 Factorising by taking out common factors; **16+:** 6.4 Factorise expressions

Key points

* **Factorising** is the opposite **operation** of expanding **brackets** so you need to put brackets in.
* To factorise an expression, you need to find any factors common to all the terms and put the common factor outside the bracket.
* You then need to decide what to put inside the bracket, through using multiplication.
* Check your answer by expanding the brackets.

Guided

1 Factorise

a $10x + 15$
$= 5(2x + \ldots)$
Check: $5(2x + \ldots) = 10x + \ldots\ldots$

b $2 - 12a$
$= 2(1 - \ldots)$
Check: $2(1 - \ldots) = \ldots\ldots$

c $y^2 - 3y$
$= y(\ldots\ldots)$
Check: $y(\ldots\ldots) = \ldots\ldots$

2 Factorise

a $3k^2 - 27k$

$= 3k(k - \ldots)$

> **Remember this**
> You can tell $3k^2 - 27k = 3(k^2 - 9k)$ is not fully factorised, because k is a common factor inside the bracket.

b $14b + 21b^2$

$= 7\ldots(2 + \ldots)$

c $8xy^2 - 16x^2y$

$= 8xy(\ldots\ldots)$

> **Remember this**
> In factorising questions, you must check that you have taken out **all** the common factors, otherwise the expression is not 'fully' factorised.

Practice

3 Factorise

a $8b - 12$

b $3 + 9p$

c $6h - 5h^2$

4 Factorise

a $2xy - 8x^2$

b $12g^2 + 18g$

c $10ab - 5a^2b + 15ab^2$

Step into GCSE

5 Factorise

a $6x^3 - 9x^2$

b $4ab + 8bc - 2bd$

c $12pq^2 - 16p^2q + 4pq$

Needs more practice ☐ Almost there ☐ I'm proficient! ☐

2.4 Laws of indices

By the end of this section you will know how to:

* Manipulate powers of a variable

GCSE LINKS
AH: 1.5 Understanding the index laws, 2.3 Using the index laws; **BH:** Unit 2 1.4 Understanding the index laws, 7.3 Using the index laws; **16+:** Use powers in algebra

Key points

* In the expression x^n, the number n is called the **power** or **index**.
* Here are three of the laws for indices:
  - $x^m \times x^n = x^{m+n}$
  - $x^m \div x^n = x^{m-n}$
  - $(x^m)^n = x^{mn}$

Guided

1 Simplify

a $y^2 \times y^4$

$= y^{2+4}$

$= y^{\dots\dots}$

Hint

$y^2 \times y^4$
$= (y \times y) \times (y \times y \times y \times y)$

b $x^5 \div x^3$

$= \frac{x^5}{x^3}$

$= x^{5-3}$

$= x^{\dots\dots}$

Hint

$\frac{x^5}{x^3}$

$= \frac{{}^1\cancel{x} \times {}^1\cancel{x} \times {}^1\cancel{x} \times x \times x}{{}_1\cancel{x} \times {}_1\cancel{x} \times {}_1\cancel{x}}$

$= \frac{x \times x}{1}$

c $x^3 \div x^3$

$= \frac{x^3}{x^3}$

$= \frac{{}^1\cancel{x} \times {}^1\cancel{x} \times {}^1\cancel{x}}{{}_1\cancel{x} \times {}_1\cancel{x} \times {}_1\cancel{x}}$

$= \frac{1}{1}$

$= \dots\dots$

Remember this

In Q1c, using the second law of indices,
$x^3 \div x^3 = x^{3-3} = x^0$,
so $x^0 = 1$.

2 Simplify

a $(a^2)^3$

$= a^2 \times a^2 \times a^2$

$= a^{2+2+2}$

$= a^{2 \times 3}$

$= a^{\dots\dots}$

Remember this

In Q2a, you do not need to write out the powers in full, but can use the third law.

b $(b^3)^4$

$= b^{3 \times 4}$

$= b^{\dots\dots}$

c $(p^4)^5$

$= p^{\dots\dots}$

$= p^{\dots\dots}$

Practice

3 Simplify

a $x^2 \times x^3$

b $y^5 \times y^7$

c $p^5 \div p^3$

d $b^6 \div b^5$

e $(r^6)^4$

f $(m^5)^3$

4 Simplify

a $a^4 \times a^9$

b $k^{10} \times k$

Remember this

$k = k^1$

c $m^{11} \div m^5$

d $b^8 \div b$

e $(k^3)^4$

f $(r^7)^3$

Guided

5 Simplify

a $2c^4 \times 3c^5$

$= 2 \times 3 \times c^4 \times c^5$

$= \dots\dots \times c^{\dots\dots}$

$= \dots\dots c^{\dots\dots}$

b $\frac{10x^7}{5x^4}$

$= \frac{10}{5} \times \frac{x^7}{x^4}$

$= 2 \times x^{7-4}$

$= \dots\dots \times x^{\dots\dots}$

$= \dots\dots x^{\dots\dots}$

Practice

6 Simplify

a $4d^5 \times 6d^7$

b $\frac{12y^6}{3y^2}$

c $5k^6 \times 3k^7$

d $\frac{27b^9}{9b^2}$

Guided

7 Simplify

a $4a^3b^5 \times 7a^4b^6$

$= 4 \times 7 \times a^3 \times a^4 \times b^5 \times b^6$

$= \ldots\ldots \times a^{\ldots\ldots} \times b^{\ldots\ldots}$

$= \ldots\ldots\ldots\ldots\ldots\ldots$

b $3c^2d^3 \times 12c^7d^8$

$= 3 \times \ldots\ldots \times c^{\ldots\ldots} \times c^{\ldots\ldots} \times d^{\ldots\ldots} \times d^{\ldots\ldots}$

$= \ldots\ldots \times c^{\ldots\ldots} \times d^{\ldots\ldots}$

$= \ldots\ldots\ldots\ldots\ldots\ldots$

c $(2a^3)^4$

$= 2a^3 \times 2a^3 \times 2a^3 \times 2a^3$

$= 2^4 \times (a^3)^4$

$= \ldots\ldots \times a^{\ldots\ldots}$

$= \ldots\ldots\ldots\ldots\ldots\ldots$

d $(3b^5)^2$

$= 3^{\ldots\ldots} \times (b^{\ldots\ldots})^{\ldots\ldots}$

$= \ldots\ldots \times b^{\ldots\ldots}$

$= \ldots\ldots\ldots\ldots\ldots\ldots$

Practice

8 Simplify

a $6x^3y^4 \times 5x^5y$

b $8p^5q^2 \times 4p^3q^6$

c $(2x^4)^5$

d $(3y^6)^4$

Step into GCSE

9 Simplify

a $5p^2q \times 3pq \times 2pq^2$

b $(4x^2y^3)^4$

c $\frac{81a^3b^2}{27ab^3}$

Don't forget!

* Simplify the following:

$a + a + a + a =$ .... $a \times a \times a \times a =$ ....

$3k - 2j + 4k - 5j =$ .... $2r^2 + 6r - 7 + 3r^2 - 9r - 5 =$ ....

* Complete the following:

$3(5t - 7) = 15t$ .... $18s + 3s^2 = 3s($ .... $)$ $x^m \times x^n = x^{\dots\dots}$

$x^m \div x^n = x^{\dots\dots}$ $(x^m)^n = x^{\dots\dots}$ $x^{\dots\dots} = 1$

$x^{\dots\dots} = x$ $p^4 \times p^5 = p^{\dots\dots}$ $p^9 \div p^5 = p^{\dots\dots}$

$(b^3)^5 = b^{\dots\dots}$ $3a^2b^7 \times 5a^3b^2 = 15$ .... $\frac{4y^3}{2y} =$ ....

Exam-style questions

1 **a** Simplify $5k - 3k^2 + 7 - 2k + 6k^2 - 9$

....

b Simplify $x^5 \times x^4$

....

c Simplify $s^9 \div s^3$

....

d Simplify $(g^7)^3$

....

2 **a** Expand $4(5r - 6s)$

....

b Expand $3g(2g^2 - 7g + 8)$

....

3 **a** Expand and simplify $3(5h - 7) - 2(6h + 4)$

....

b Expand and simplify $x(4x + 5) + 3x(2x - 9)$

....

4 **a** Factorise $6b - 9a + 3$

....

b Factorise $4x^3 + 12x^2 - 6x$

....

c Factorise $5x^2y - 10xy^2 + 15xy$

....

5 **a** Simplify $7a^2 \times 3a^3$

b Simplify $\frac{14p^7}{2p^4}$

c Simplify $5v^3w^2 \times 4v^4w^5$

d Simplify $(3g^4)^4$

6 **a** Simplify $4h^2 - 7h + 5 - h^2 + 5h - 8$

b Simplify $5g^4 \times 3g^5$

c Simplify $\frac{18k^5}{3k^2}$

d Simplify $(2m^3)^5$

7 **a** Expand $2xy(3x - y)$

b Expand and simplify $2k(3k - 8) - 4k(5k + 6)$

8 **a** Factorise $3x - 5x^2 + 7x^3$

b Factorise $2vw^2 - 4v^2w + 8v^2w^2$

9 **a** Expand $n^3(n^4 - n^2)$

b Expand $p^2(p^5 + p^4 - p^3)$

3.1 Using word formulae

By the end of this section you will know how to:

* Substitute numbers into a word formula

GCSE LINKS

AF: 28.1 Using word formulae; **BF:** Unit 2 13.1 Using word formulae; **AH:** 19.5 Using formulae; **BH:** Unit 2 10.2 Using formulae

Key points

* A **formula** is a mathematical relationship or rule expressed in words or symbols.
* A **word formula** uses words to show the relationship. For example, for a rectangle: area = length × width.
* A word formula can be rearranged to find the value of any of the **variables**.

Remember this

Formulae is the plural of formula.

Remember this

A **variable** is a value that can change.

Guided

1 This word formula can be used to work out the perimeter of a square:

perimeter = 4 × length of side

Work out the perimeter of a square with sides of length:

a 3 cm

Perimeter = 4 × 3

Perimeter = ... cm

b 5 cm

Perimeter = 4 × ....

Perimeter = ... cm

The perimeter of a square is 40 cm.

c Work out the length of each side.

40 = 4 × length of side

40 ÷ .... = length of side

.... cm = length of side

2 The total cost of hiring a car can be worked out using this formula:

total cost = cost per day × number of days

Work out the total cost of hiring a car for:

a 4 days at £12 per day

Total cost = £12 × 4

Total cost = £........

b 5 days at £11 per day

Total cost = £11 × ....

Total cost = £........

The total cost of hiring a car for 4 days is £60.

c What is the cost per day?

£60 = cost per day × 4

£60 ÷ .... = cost per day

£........ = cost per day

Practice

3 The average speed of a car in kilometres per hour can be worked out using the formula:

$$\text{average speed} = \frac{\text{total distance in kilometres}}{\text{total time in hours}}$$

a Work out the average speed of a car which travels 120 kilometres in 3 hours.

.... kilometres per hour

b Work out the total distance travelled by a car which travels at an average speed of 100 kilometres per hour for 2 hours.

.... kilometres

c Work out the time taken for a car to travel 160 kilometres at an average speed of 80 kilometres per hour.

.... hours

4 This formula can be used to work out the cooking time for roast beef:

time in minutes = 15 × weight in pounds + 15

a Work out the cooking time for a piece of beef that weighs 4 pounds.

.... minutes

b A piece of beef takes 90 minutes to cook. Work out the weight of the piece of beef.

.... pounds

5 The formula to work out the area of a triangle is:

$$\text{area} = \frac{1}{2} \times \text{base} \times \text{vertical height}$$

Work out the area of a triangle with base 6 cm and height 8 cm.

.... cm^2

Step into GCSE

6 The formula to work out the weekly pay of a shop assistant is:

total pay = hourly rate of pay × number of hours worked − deductions

Work out the total pay of a shop assistant who works for 20 hours at £8 per hour with deductions of £40.

£ ....

7 This formula can be used to work out a monthly phone bill:

total bill = cost per minute × number of minutes + monthly charge

Work out the total bill in a month when the monthly charge is £15 and 100 minutes of calls are made at a cost of 6p per minute.

Remember this

If a question involves mixed units, make sure you convert some so you work in consistent units.

£ ....

Needs more practice ☐ Almost there ☐ I'm proficient! 

3.2 Using algebraic formulae

By the end of this section you will know how to:

- Substitute numbers into an algebraic formula

GCSE LINKS

AH: 19.5 Using formulae; **BH:** Unit 2 10.2 Using formulae; **16+:** 10.1 Use algebraic formulae

Key points

- A **formula** is a mathematical relationship or rule expressed in words or symbols (letters).
- An **algebraic** formula uses letters to show the relationship, for example, $d = st$.
- The letter on its own on one side of the = sign is called the **subject** of the formula; in the formula $d = st$, d is the subject.
- You can find the value of a letter that is not the **subject** of an algebraic formula.

Guided

1 Use the formula $s = 5t$ to:

You should know

$5t$ stands for $5 \times t$

a find the value of s when $t = 2$

$s = 5 \times 2$

$s =$ ....

b find the value of s when $t = -4$

$s = 5 \times$ ....

$s =$ ....

c find the value of t when $s = 15$

$15 = 5t$

$t = \frac{15}{\ldots\ldots}$

$t =$ ....

2 Use the formula $y = 3x - 5$ to:

a find the value of y when $x = 6$

$y = 3 \times 6 - 5$

$y =$ ....

b find the value of y when $x = -2$

$y = 3 \times$ .... $- 5$

$y =$ ....

c find the value of y when $x = \frac{1}{3}$

$y =$ ....

$y =$ ....

Remember this

When substituting values in formulae you need to follow the BIDMAS rule.

d find the value of x when $y = 13$

$13 = 3x - 5$

$13 +$ .... $= 3x$

.... $= 3x$

.... $\div$ .... $= x$

$x =$ ....

Practice

3 Use the formula $T = 30W + 20$ to:

a find the value of T when $W = 10$

b find the value of T when $W = -8$

c find the value of T when $W = 0.5$

d find the value of W when $T = 50$

4 Use the formula $V = lwh$ to find the value of V when $l = 3$, $w = 4$ and $h = 2$.

5 Use the formula $s = \frac{d}{t}$ to find the value of s when $d = 10$ and $t = 2$.

Remember this

$\frac{d}{t}$ is the same as $d \div t$

6 Use the formula $v = u - gt$ to find the value of v when:

a $u = 12$, $g = 10$ and $t = 1$

b $u = 25$, $g = -10$ and $t = 2$

7 Use the formula $s = \frac{u + v}{2} \times t$ to find the value of s when:

a $u = 0$, $v = 8$ and $t = 2$

b $u = 3$, $v = 5$ and $t = 4$

8 Use the formula $s = ut + \frac{1}{2}at^2$ to find the value of s when:

a $u = 0$, $a = 4$ and $t = 2$

b $u = 1$, $a = -2$ and $t = 3$

Step into GCSE

9 Use the formula $v^2 = u^2 + 2as$ to find the value of v when $u = 2$, $a = 2$ and $s = 3$.

10 Use the formula $u = \sqrt{v^2 - 2as}$ to find the value of u when $v = 5$, $a = 4$ and $s = 2$.

Needs more practice ☐ Almost there ☐ I'm proficient! ☐

3.3 Changing the subject of a formula

By the end of this section you will know how to:

* Change the subject of a formula where the subject only appears once

GCSE LINKS
AH: 19.7 Changing the subject of a formula; **BH:** Unit 3 5.5 Changing the subject of a formula; **16+:** 10.4 Change the subject of a formula

Key points

* The letter on its own on one side of the = sign is called the **subject** of the formula.
* You can change the subject of a formula by:
  - using a reverse flowchart, or
  - carrying out the same operations on both sides of the equals sign.

Guided

1 Make x the subject of the formula $y = 4x - 5$.

Flowchart method

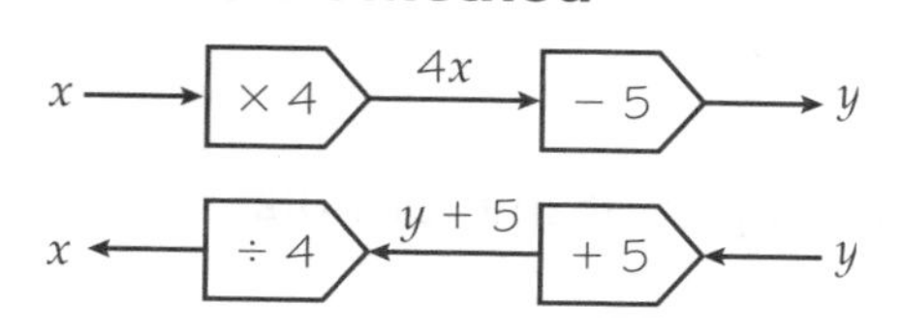

Remember this
First draw a flowchart for the formula, starting with the letter that you want to make the subject. Reverse the flowchart to rearrange the formula.

You should know
To 'undo' the formula you need to do the opposite operation, so change + to − (and − to +), and change × to ÷ (and ÷ to ×).

So $x = \frac{y + \text{.........}}{\text{.........}}$ or $x = \frac{1}{4}(y + \text{.........})$

Algebraic method

$y = 4x - 5$

$y + 5 = 4x$ add 5 to both sides

$\frac{\text{...............}}{\text{...........}} = x$ divide both sides by 4

Answer: $x = \frac{\text{...............}}{\text{...........}}$

Practice

2 Make W the subject of the formula $T = 30W + 20$.

3 Make Q the subject of the formula $R = \frac{Q}{7}$.

4 Make k the subject of the formula $P = \frac{k}{4} - 5$.

5 Make s the subject of the formula $V = \frac{3}{s}$.

6 Make R the subject of the formula $L = \frac{R}{5} + 2$.

7 Make t the subject of the formula $v = u + at$.

8 Make d the subject of the formula
$s = \frac{d}{t}$.

9 Make w the subject of the formula
$V = lwh$.

Remember this

First, either multiply out the brackets or divide each side by 2.

10 Make l the subject of the formula
$P = 2(l + w)$.

11 Make T the subject of the formula
$I = \frac{PRT}{100}$.

12 Make V the subject of the formula
$D = \frac{M}{V}$.

13 Make a the subject of the formula
$P = 2a + 3b^2$.

14 Make r the subject of the formula
$A = \pi r^2$.

Step into GCSE

Don't forget!

* Complete the following:

A .... is a mathematical relationship or rule expressed in .... or ....

* This word formula can be used to work out the perimeter of an equilateral triangle:

perimeter = 3 × length of side

Work out the perimeter of an equilateral triangle with side of 5 cm.

* Use the formula $P = 2a + b$ to find the value of P when $a = 3$ and $b = 4$

* Use the formula $V = \frac{E}{Q}$ to find the value of V when $E = 15$ and $Q = 3$

* Complete the following:

If $W = VA$ then $V = \frac{\dots}{\dots}$

If $P = \frac{W}{T}$ then $W =$ ....

If $P = \frac{F}{A}$ then $A = \frac{\dots}{\dots}$

If $P = 2x + 3y$ then $x = \frac{\dots}{\dots}$

Exam-style questions

1 Here is a word formula a plumber uses to work out the total cost of a job:

Total cost = cost per hour × number of hours + call-out charge

The cost per hour is £20 and the call-out charge is £50

a Work out the total cost of a job that takes him 2 hours.

£ ....

b The plumber charges £150 for a job. How many hours did the job take?

.... hours

2 A shop sells T-shirts which can have names printed on them. Here is the word formula the shop uses to work out the total cost of a T-shirt with a printed name:

Total cost = cost of T-shirt + cost per letter × number of letters

What is the total cost of a T-shirt which costs £15 and has the name DANNY printed at a cost of 10p per letter?

£ ....

3 Here is a formula $P = 3x - 5$

a Find the value of P when $x = 10$

....

b Find the value of x when $P = 16$

....

c Make x the subject of the formula.

....

4 Here is a formula $W = m(v - u)$

a Find the value of W when $m = 10$, $v = 8$ and $u = 3$

....

b Find the value of W when $m = 0.5$, $v = 10$ and $u = -2$

....

5 Here is the formula for the area, A, of a triangle: $A = \frac{bh}{2}$

a Find the value of A when $b = 3$ and $h = 4$

....

b Make h the subject of the formula.

....

6 Here is a formula $P = 2a + 3b$

a Find the value of P when $a = 2$ and $b = 3$

....

b Find the value of a when $P = 20$ and $b = 4$

....

c Make b the subject of the formula.

....

7 Here is a formula $v = u - gt$

a Find the value of v when $u = 25$, $g = 10$ and $t = 1$

....

b Find the value of v when $u = 4$, $g = -10$ and $t = 2$

....

8 Make m the subject of the formula $E = \frac{1}{2}mv^2$

....

9 Make c the subject of the formula $s = \frac{a + b + c}{2}$

....

10 Make b the subject of the formula $S = 2(3a + 2b)$

....

11 Make h the subject of the formula $A = \frac{1}{2}(a + b)h$

....

Needs more practice ☐ Almost there ☐ I'm proficient! ☐

4.1 Solving equations with one operation

GCSE LINKS
AH: 13.1 Solving simple equations; **BH:** Unit 3 4.1 Solving simple equations; **16+:** 9.1 Solve equations with variables on one side

By the end of this section you will know how to:

* Solve linear equations with one operation

Key points

* A **linear** equation is one which does not involve any powers.
* The graph of a linear equation is a straight line. (See Chapters 7 and 8 for more information.)
* In an equation, the left-hand side must always be equal to the right-hand side.
* You can solve an equation by rearranging it and applying suitable operations (+, −, ×, ÷) to get the **unknown** on its own on one side of the equation.
* You must apply the same operation to both the left-hand and right-hand sides of the equation; this is called the **balance method**.

Remember this
To solve means to find the value of the unknown.

Remember this
The letter whose value you need to work out is called the unknown or variable.

Guided

1 Solve

a $x + 5 = 8$

$x + 5 - 5 = 8 - 5$

$x =$ ....

Remember this
For $x + 5 = 8$, to get x on its own you need to undo $+ 5$, so you do the opposite operation (i.e. $- 5$) to both sides of the equation.

b $y - 7 = 3$

$y - 7 + 7 = 3 +$ ....

$y =$ ....

Remember this
You can check your answer by substituting the value back into the original equation. For example, if your answer to $y - 7 = 3$ is $y = 10$, checking by substitution gives: $10 - 7 = 3$ ✓

c $9 = 2 + z$

$9 - 2 = 2 + z - 2$

.... $= z$ or $z =$ ....

d $p + 8 = 2$

$p + 8 - 8 = 2 - 8$

$p =$ ....

Practice

2 Solve

a $s + 10 = 15$

b $t - 9 = 1$

c $7 + k = 5$

d $11 = p - 8$

e $10 = 10 + w$

f $v - 1 = 20$

Guided

3 Solve

a $4x = 20$

$\frac{4x}{4} = \frac{20}{4}$

$x =$ ....

You should know
+ and − are opposite operations; × and ÷ are opposite operations.

b $\frac{1}{3}y = 4$

Remember this
$\frac{1}{3}y = \frac{y}{3} = y \div 3$

$\frac{1}{3}y \times 3 = 4 \times 3$

$y =$ ....

Remember this
You can check your answer by substituting the value back into the original equation. For example, if $\frac{1}{3}y = 4$ gives $y = 12$, checking by substitution gives: $\frac{1}{3} \times 12 = 4$ ✓

c $5 = 25k$

$25k = 5$

$\frac{25k}{\ldots\ldots} = \frac{5}{\ldots\ldots}$

$k = \frac{\ldots\ldots}{\ldots\ldots}$

Practice

4 Solve

a $3k = 15$ **b** $\frac{1}{4}g = 7$ **c** $5w = 45$

d $\frac{1}{6}v = 3$ **e** $6s = 24$ **f** $\frac{1}{10}t = 4$

g $9 = 27c$ **h** $10 = 100x$ **i** $0.5y = 12$

Step into GCSE

5 Solve $\frac{2}{3}x = 12$

Hint
To solve $\frac{2}{3}x = 12$, first get rid of the fraction, by multiplying both sides by 3.

Needs more practice ☐ Almost there ☐ I'm proficient! ☐

4.2 Solving equations with two operations

GCSE LINKS
AH: 13.1 Solving simple equations; **BH:** Unit 3 4.1 Solving simple equations; **16+:** 9.1 Solve equations with variables on one side

By the end of this section you will know how to:

* Solve linear equations with two operations

Key points

* In an equation with two operations, deal with any + or − first.
* Solutions to equations can be whole numbers, fractions, mixed numbers, improper fractions or decimals.
* Solutions to linear equations can be negative.

Guided

1 Solve

a $4a + 3 = 11$

$4a + 3 - 3 = 11 - 3$

$4a =$ ....

$\frac{4 \times a}{4} = \frac{.........}{4}$

$a =$ ....

b $3x - 7 = 8$

$3x - 7 + 7 = 8 + 7$

$3x =$ ....

$\frac{3 \times x}{3} = \frac{.........}{3}$

$x =$ ....

c $\frac{1}{7}s + 2 = 3$

$\frac{1}{7}s + 2 - 2 = 3 - 2$

$\frac{1}{7}s =$ ....

$\frac{s \times 7}{7} =$ .... × ....

$s =$ ....

Hint
$\frac{1}{7}s = \frac{s}{7}$

Practice

2 Solve

a $2a - 7 = 5$ **b** $5b + 1 = 46$ **c** $\frac{1}{4}k - 6 = 3$

d $\frac{1}{3}m + 4 = 7$ **e** $6a - 5 = 13$ **f** $\frac{1}{6}w - 2 = 8$

Guided

3 Solve

a $5s - 8 = 6$

$5s - 8 + 8 = 6 + 8$

$5s = \dots\dots$

$\frac{5 \times s}{5} = \frac{\dots\dots}{5}$

$s = \frac{\dots\dots}{5}$

Remember this

The answer can be left as an improper fraction. For example, $\frac{14}{5}$ is acceptable, as well as $2\frac{4}{5}$ or 2.8

b $3k + 10 = 4$

$3k + 10 - 10 = 4 - 10$

$3k = - \dots\dots$

$\frac{3 \times k}{3} = - \frac{\dots\dots}{3}$

$k = - \dots\dots$

You should know

For multiplication and division involving negative numbers, if the signs are the same the result is positive, but if the signs are different the result is negative.

Practice

4 Solve

a $3v + 6 = 7$ **b** $7p + 11 = 4$ **c** $2w + 3 = -7$

d $8k - 1 = 12$ **e** $6x + 5 = 3$ **f** $5y - 7 = -9$

GCSE

5 In this quadrilateral, the sizes of the angles are $2x$, $3x - 10°$, $x + 50°$ and $80°$.

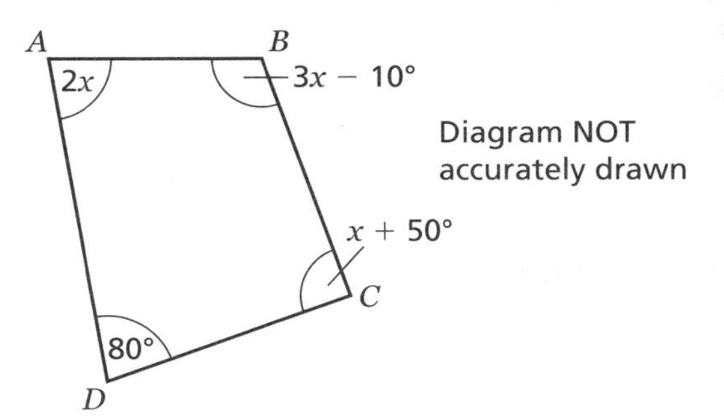

a Use this information to write down an equation in x.

b Work out the value of x.

Hint

The sum of the angles of a quadrilateral is 360°. Simplify your equation in Q5a by collecting like terms before solving.

Needs more practice ☐ Almost there ☐ I'm proficient! ☐

4.3 Solving equations with brackets

By the end of this section you will know how to:

* Solve equations involving brackets

GCSE LINKS
AH: 13.2 Solving linear equations containing brackets; **BH:** Unit 3 4.2 Solving linear equations involving brackets; **16+:** 9.2 Solve equations with brackets

Key points

* In an equation with brackets, expand the brackets first.

Guided

1 Solve

a $4(x - 1) = 5$

$4x - 4 = 5$

$4x - 4 + \dots = 5 + \dots$

$4x = \dots$

$\frac{4x}{\dots} = \frac{\dots}{\dots}$

$x = \frac{\dots}{\dots}$

b $2(4w + 13) = 20$

$8w + \dots = 20$

$8w + \dots - \dots = 20 - \dots$

$8w = \dots$

$\frac{8w}{\dots} = \frac{\dots}{\dots}$

$w = \frac{\dots}{\dots}$

c $3(5k - 2) - 4 = 2$

$15k - 6 - 4 = 2$

$15k - \dots = 2$

$15k - \dots + \dots = 2 + \dots$

$15k = \dots$

$\frac{15k}{\dots} = \frac{\dots}{\dots}$

$k = \frac{\dots}{\dots}$

d $5 = 3(2p + 1)$

$5 = 6p + 3$

$6p + 3 = 5$

$6p + 3 - \dots = 5 - \dots$

$6p = \dots$

$\frac{6p}{\dots} = \frac{\dots}{\dots}$

$p = \frac{\dots}{\dots}$

Remember this
If the term with the unknown is on the right-hand side, first re-write the equation so that the unknown is on the left-hand side before solving. For example, re-write $3 = -4x$ as $-4x = 3$ and so $4x = -3$.

Practice

2 Solve

a $4(y - 3) = 18$

b $3(p + 7) = 11$

c $2(3r - 5) = 14$

d $5(2k + 1) = 3$

e $10 = 6(4w - 2)$

f $2(x + 7) + 5 = 3$

g $4(2y - 3) - 1 = 11$

h $5 - (3x + 2) = 8$

i $6(5 + 2y) - 8 = 10$

Guided

3 Solve $\frac{w-3}{2} = 4$

$\frac{(w-3) \times 2}{2} = 4 \times 2$

$(w-3) =$ ....

$w - 3 =$ ....

$w - 3 +$ .... $=$ .... $+$ ....

$w =$ ....

Remember this

When solving equations involving algebraic fractions, first put in brackets on the top line (numerator). For example, $\frac{(w-3)}{2} = 4$ shows that all of $w - 3$ is divided by 2.

Remember this

$\frac{w-3}{2} = \frac{(w-3)}{2}$

$= \frac{1}{2}(w-3)$

Practice

4 Solve

a $\frac{x+5}{3} = 7$

b $\frac{y-4}{5} = 2$

c $5 = \frac{2k+9}{4}$

GCSE

5 Solve

a $4 + 2(3p - 1) = 10$

b $\frac{5w-3}{3} = 9$

Needs more practice ☐ Almost there ☐ I'm proficient! ☐

4.4 Solving equations where the unknown appears on both sides

By the end of this section you will know how to:

* Solve equations where the unknown appears on both sides

GCSE LINKS

AH: 13.3 Solving linear equations with the unknown on both sides; **BH:** Unit 3 4.3 Solving linear equations with the unknown on both sides; **16+:** 9.3 Solve equations with variables on both sides

Key points

* Collect like terms so that the terms involving the unknown are on one side of the equation and everything else is on the other side.

Remember this

Take the terms involving the unknown to the side where there are more of them.

Remember this

There is not only one way to approach solving equations. No operation is wrong as long as it is done correctly to both sides. Some operations may be more suitable than others.

Guided

1 Solve

a $5a - 4 = 3a + 6$

Method 1

$5a - 4 = 3a + 6$

$5a - 4 + 4 = 3a + 6 + 4$

$5a = 3a + 10$

$5a - 3a = 3a + 10 - 3a$

$\ldots\ldots a = 10$

$a = \ldots\ldots$

b $4b + 10 = b + 1$

Method 1

$4b + 10 = b + 1$

$4b + 10 - 10 = b + 1 - 10$

$4b = b - \ldots\ldots$

$4b - b = b - \ldots\ldots - b$

$\ldots\ldots b = -\ldots\ldots$

$b = -\ldots\ldots$

c $3c + 5 = 6c - 7$

Method 1

$3c + 5 = 6c - 7$

$3c + 5 + 7 = 6c - 7 + 7$

$3c + \ldots\ldots = 6c$

$3c + \ldots\ldots - 3c = 6c - 3c$

$\ldots\ldots = \ldots\ldots c$

$\ldots\ldots c = \ldots\ldots$

$c = \ldots\ldots$

Remember this

When solving $3c + 5 = 6c - 7$ take the unknowns to the right-hand side as $6c$ is bigger than $3c$, then re-write so that the term involving the unknown is on the left-hand side.

a **Method 2**

$5a - 4 = 3a + 6$

$5a - 4 - 3a = 3a + 6 - 3a$

$2a - 4 = 6$

$2a - 4 + 4 = 6 + 4$

$2a = \ldots\ldots$

$a = \ldots\ldots$

b **Method 2**

$4b + 10 = b + 1$

$4b + 10 - b = b + 1 - b$

$3b + 10 = 1$

$3b + 10 - 10 = 1 - 10$

$\ldots\ldots b = -\ldots\ldots$

$b = -\ldots\ldots$

c **Method 2**

$3c + 5 = 6c - 7$

$3c + 5 - 3c = 6c - 7 - 3c$

$5 = 3c - 7$

$3c - 7 = 5$

$3c - 7 + 7 = 5 + 7$

$\ldots\ldots c = \ldots\ldots$

$c = \ldots\ldots$

Practice

2 Solve

a $4x - 7 = 2x + 3$

b $3k + 2 = k - 6$

c $5v - 9 = 3v - 4$

d $2s + 7 = 3s - 8$

e $6 + 4h = 9 + 7h$

f $8t - 5 = 3t + 7$

Step into GCSE

3 Solve $4(3x + 2) = 3(2x - 1)$

4.5 Solving equations with negative coefficients

GCSE LINKS
AH: 13.1 Solving simple equations; **BH:** Unit 3 4.1 Solving simple equations; **16+:** 9.4 Solve equations with negative coefficients

By the end of this section you will know how to:

* Solve equations with negative coefficients

Key points

* A **coefficient** is the number in front of the unknown.
* You solve equations with negative coefficients using the balance method as before.

You should know
There may be different ways to solve a problem.

Guided

1 Solve

a $8 - 2x = 12$

Method 1

$8 - 2x = 12$

$8 - 2x + 2x = 12 + 2x$

$8 = 12 + 2x$

$8 - 12 = 12 + 2x - 12$

$-$ .... $= 2x$

$2x = -$ ....

$x = -$ ....

Remember this
One way is to add $2x$ to both sides to give a positive term in x.

Method 2

$8 - 2x = 12$

$8 - 2x - 8 = 12 - 8$

$-2x = 4$

$2x = -$ ....

$x = -$ ....

Remember this
If $-2x = 4$ then $2x = -4$

b $6 - 3y = 9 - 5y$

Remember this
In $6 - 3y = 9 - 5y$, -3 is greater than -5 so one approach is to take the ys to the left-hand side of the equation where there are more of them.

Method 1

$6 - 3y = 9 - 5y$

$6 - 3y - 6 = 9 - 5y - 6$

$-3y =$ .... $- 5y$

$-3y + 5y =$ .... $- 5y + 5y$

.... $y =$ ....

$y = \frac{.........}{.........}$

Method 2

$6 - 3y = 9 - 5y$

$6 - 3y + 3y = 9 - 5y + 3y$

$6 = 9 -$ .... y

$6 -$ .... $= 9 -$ .... $y -$ ....

$-$ .... $= -$ .... y

.... $y =$ ....

$y = \frac{.........}{.........}$

Remember this
If $-3 = -2y$ then $2y = 3$

Practice

2 Solve

a $4 - p = 3$

b $11 - 3x = 2$

c $15 = 7 - 2y$

d $8 - 4k = 6k + 5$

e $3 - 5w = 7 - w$

f $2(1 - 3s) = 7 - 4s$

3 Solve $4(1 - 2x) - 3(x + 2) = 0$

Don't forget!

* Complete the sentences below using the words in the box.

balance method	coefficient	brackets	powers	more of them
left-hand side	+ or −	right-hand side		

* A linear equation is one which does not involve any ....
* In an equation, the .... must always be equal to the ....
* You must apply the same operation to both the left-hand and right-hand sides of the equation; this is called the ....
* In an equation with two operations, deal with any .... first.
* In an equation with ...., expand the .... first.
* When the unknown is on both sides of the equation, take the terms involving the unknown to the side where there are ....
* A .... is the number in front of the unknown.
* Complete the following:

$3y = 15$ $y =$ ....

$2x + 5 = 17$ $x =$ ....

$\frac{w - 6}{3} = 4$ $w =$ ....

$15 - 2v = 7$ $v =$ ....

Exam-style questions

1 **a** Solve $3s - 7 = 5$

b Solve $2(3k - 4) = 10$

c Solve $6 - 4p = p - 4$

d Solve $2(5x + 1) = 3(2x - 1)$

2 **a** Solve $\frac{1}{3}w + 7 = 5$

b Solve $3(2w - 1) = 9$

c Solve $4 - 2v = 11 - 5v$

d Solve $\frac{y - 5}{2} = 7$

3 **a** Solve $8 - \frac{1}{2}d = 3$

b Solve $5(3 - 4t) = 7$

c Solve $6r - 7 = 4 - 2r$

d Solve $9 = \frac{2x + 3}{4}$

4 **a** Solve $3(x - 2) = 6$

b Solve $4 - 2w = 7 - 3w$

c Solve $2(1 - 3k) = 3(2 + k)$

d Solve $\frac{7x - 1}{3} = 5$

Needs more practice ☐ Almost there ☐ I'm proficient! ☐

5.1 Introducing inequalities

GCSE LINKS
AF: 21.9 Introducing inequalities;
BF: Unit 3 4.1 Introducing inequalities

By the end of this section you will know how to:

- Use symbols for inequality

Key points

- $>$ means **greater than**
- $\geqslant$ means **greater than or equal to**
- $<$ means **less than**
- $\leqslant$ means **less than or equal to.**

Remember this
If you think of the $>$ sign as an arrow, it always points towards the smaller number.

Guided

1 Put the correct inequality sign between each pair of numbers.

a 9 ..>.. 5 **b** 3 ..<.. 7 **c** 2 .... −3

Practice

2 Put the correct inequality sign between each pair of numbers.

a 0 .... 2 **b** 7 .... −5 **c** 9.2 .... 11.1

3 Write down whether each statement is true or false.

a $4 > 2$ **b** $3.5 < 2$ **c** $7 > -9$

....

Needs more practice ☐ Almost there ☐ I'm proficient! ☐

5.2 Representing inequalities on a number line

GCSE LINKS
AH: 19.1 Representing inequalities on a number line;
BH: Unit 3 5.1 Representing inequalities on a number line;
16+: 9.7 Understand and represent inequalities on a number line

By the end of this section you will know how to:

- Represent inequalities on a number line

Key points

- An open circle shows that the number **is not** included.
- A solid circle shows that the number **is** included.

Guided

1 Show these inequalities, in x, on the number line.

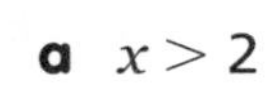

a $x > 2$

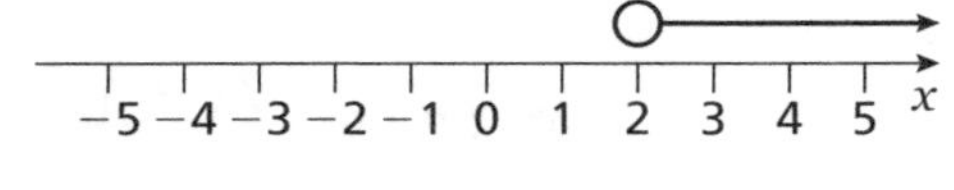

b $x \leqslant 3$

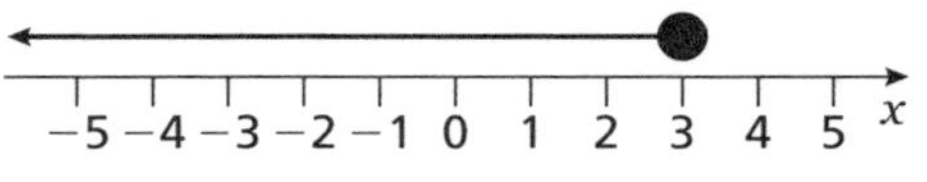

c $-1 \leqslant x < 4$

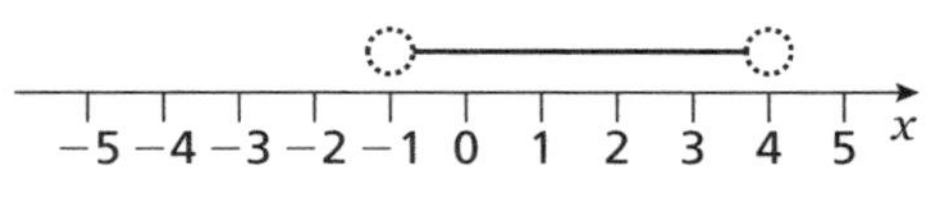

Hint
$-1 \leqslant x < 4$ means $-1 \leqslant x$, that is $x \geqslant -1$ and $x < 4$, i.e. x lies between −1 and 4, including −1 but not 4.

d $-2 < x \leqslant 1$

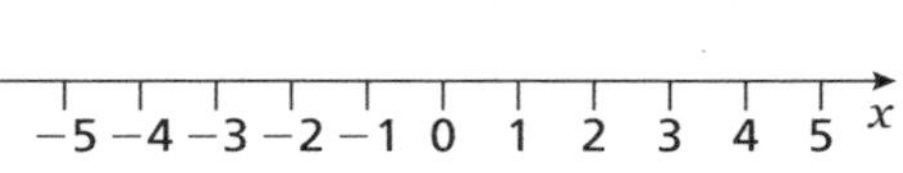

Practice

2 Show these inequalities, in x, on the number line.

a $x < 6$

−2 −1 0 1 2 3 4 5 6 7 8 x

b $x \geqslant 4$

−2 −1 0 1 2 3 4 5 6 7 8 x

c $x \leqslant 5$

−2 −1 0 1 2 3 4 5 6 7 8 x

d $x > -1$

−2 −1 0 1 2 3 4 5 6 7 8 x

3 Show these inequalities, in x, on the number line.

a $-3 < x \leqslant 2$

−5 −4 −3 −2 −1 0 1 2 3 4 5 x

b $-2 \leqslant x < 3$

−5 −4 −3 −2 −1 0 1 2 3 4 5 x

c $-4 < x < 1$

−5 −4 −3 −2 −1 0 1 2 3 4 5 x

d $-1 \leqslant x \leqslant 4$

−5 −4 −3 −2 −1 0 1 2 3 4 5 x

Guided

4 Write down the inequalities, in x, shown on the number lines.

a

−5 −4 −3 −2 −1 0 1 2 3 4 5 x

$x < 2$

b

−5 −4 −3 −2 −1 0 1 2 3 4 5 x

x .... -1

c

−5 −4 −3 −2 −1 0 1 2 3 4 5 x

$-4 \leqslant x$ 3

d

−5 −4 −3 −2 −1 0 1 2 3 4 5 x

-1 .... x 5

Practice

5 Write down the inequalities, in x, shown on the number lines.

a

0 1 2 3 4 5 6 7 8 9 10 x

....

b

0 1 2 3 4 5 6 7 8 9 10 x

....

c

0 1 2 3 4 5 6 7 8 9 10 x

....

6 Write down the inequalities, in x, shown on the number lines.

a

−5 −4 −3 −2 −1 0 1 2 3 4 5 x

....

b

−5 −4 −3 −2 −1 0 1 2 3 4 5 x

....

c

−5 −4 −3 −2 −1 0 1 2 3 4 5 x

....

Needs more practice ☐ Almost there ☐ I'm proficient! ☐

5.3 Solving linear inequalities

By the end of this section you will know how to:

* Solve linear inequalities

GCSE LINKS
AH: 19.2 Solving simple linear inequalities in one variable; **BH:** Unit 3 5.2 Solving simple linear inequalities in one variable; **16+:** 9.8 Solve inequalities

Key points

* You can solve inequalities in a similar way to linear equations. (See Chapter 4.)

Remember this
Rearrange with the unknown on the left-hand side. Also, if, for example, $5 > x$ then $x < 5$

Guided

1 Solve these inequalities.

a $x - 5 < 7$

$x - 5 + 5 < 7 + 5$

$x <$ ....

You should know
Get rid of + or − terms first.

b $2x - 3 \geqslant 5$

$2x - 3 + 3 \geqslant 5 + 3$

$2x \geqslant$ ....

$x \geqslant$ ....

c $8 > 5x - 10$

$5x - 10 < 8$

$5x - 10 + 10 < 8 + 10$

$5x <$ ....

$x < \frac{\dots\dots\dots}{\dots\dots\dots}$

2 Solve these inequalities.

a $-x > 4$

$-x + x > 4 + x$

$0 - 4 > 4 + x - 4$

$-4 > x$

$x < -4$

b $-2x < -6$

$-2x + 2x < -6 + 2x$

$0 + 6 < -6 + 2x + 6$

$6 <$ ....

$2x > 6$

$x >$ ....

c $9 - 4x \leqslant 6$

$9 - 9 - 4x \leqslant 6 - 9$

$-4x \leqslant -3$

$\frac{-4x}{-4}$ ..... $\frac{-3}{\dots\dots\dots}$

x ..... $\frac{\dots\dots\dots}{\dots\dots\dots}$

Remember this
As $-4x \leqslant -3$ you need to divide both sides by -4, which means the direction of the inequality changes.

Practice

3 Solve these inequalities.

a $4x > 16$

b $5x - 7 \leqslant 3$

c $1 \geqslant 3x + 4$

d $5 - 2x < 12$

e $\frac{x}{2} \geqslant 5$

f $8 < 3 - \frac{x}{3}$

4 Solve these inequalities.

a $\frac{x}{5} < -4$

b $10 \geqslant 2x + 3$

c $7 - 3x > -5$

GCSE

5 Solve these inequalities.

a $3t + 1 < t + 6$

....

b $2(3n - 1) \geqslant n + 5$

....

Needs more practice ☐ Almost there ☐ I'm proficient! 

5.4 Finding integer solutions to inequalities

By the end of this section you will know how to:

* Find integer solutions to inequalities

GCSE LINKS

AH: 19.3 Finding integer solutions to inequalities in one variable; **BH:** Unit 3 5.3 Finding integer solutions to inequalities in one variable; **16+:** 9.8 Solve inequalities

Key points

* An **integer** is a whole number.
* You can use a number line to help work out the integer solutions of an inequality.

Guided

1 Find the possible integer values of x in these inequalities.

Remember this

$-4 < x \leqslant 3$ means $-4 < x$, so -4 is not a solution, but $x \leqslant 3$ so 3 is a solution.

You should know

The integer values are all the whole numbers covered in the inequality.

a $-4 < x \leqslant 3$

−5 −4 −3 −2 −1 0 1 2 3 4 5

Answer: −3, −2, −1, 0, 1, 2, 3

b $-1 \leqslant x < 4$

−5 −4 −3 −2 −1 0 1 2 3 4 5

Answer: −1, 0, ...

c $0 < x \leqslant 6$

−4 −3 −2 −1 0 1 2 3 4 5 6

Answer: ....

2 Find the smallest possible integer value of x that satisfies the inequality $2x + 5 > 12$.

$2x + 5 > 12$

$2x + 5 - 5 > 12 - 5$

$2x >$ ....

$x >$ ....

Smallest integer value = ...

3 Find the possible integer values of x in these inequalities.

a $-5 \leqslant x \leqslant 4$

b $-5 < x \leqslant 0$

c $2 < x < 8$

d $-3 \leqslant x < 6$

4 Find the largest possible integer value of x that satisfies the inequality $3x - 2 \leqslant 9$.

5 Find the possible integer values of x if $-5 < 2x \leqslant 6$.

Hint

In Q5 divide all three terms by 2 to find the values x lies between.

....

Don't forget!

* Complete the sentences below using the words in the box.

less than	greater than	less than or equal to	greater than or equal to
is not	is	whole number	

- $>$ means ....
- $\geqslant$ means ....
- $<$ means ....
- $\leqslant$ means ....

* When representing inequalities on a number line:
  - an open circle shows that the number .... included
  - a solid circle shows that the number .... included.

* An integer is a ....

Exam-style questions

1 **a** Write down the inequality, in x, represented on this number line.

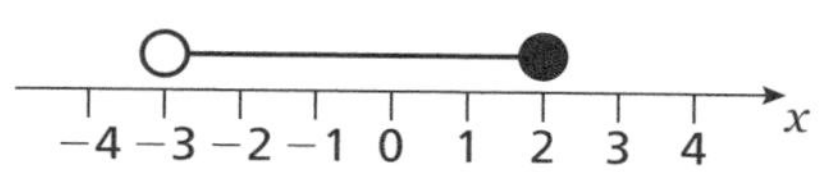

b On the number line below, show the inequality $-4 \leqslant x < 2$

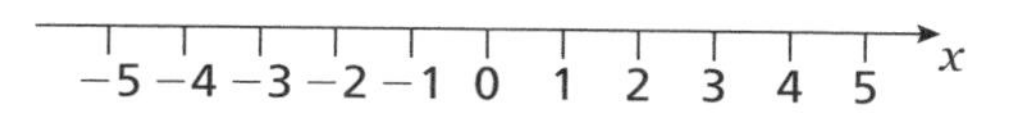

c List the integers that satisfy the inequality $-4 < x \leqslant 3$

d Solve the inequality $2t + 5 \leqslant 12$

2 **a** Write down the inequality, in y, represented on this number line.

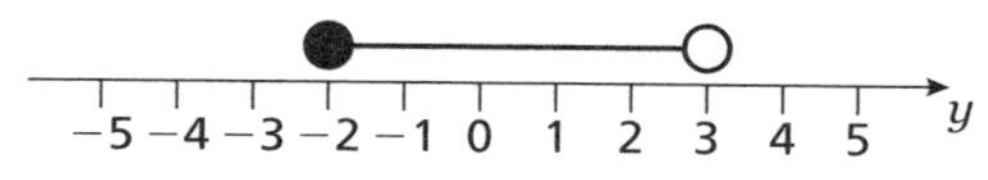

b On the number line below, show the inequality $-3 < x \leqslant 4$

−5 −4 −3 −2 −1 0 1 2 3 4 5 x

c List the integers that satisfy the inequality $-5 \leqslant x < 1$

d Solve the inequality $3x - 7 > 13$

3 **a** Write down the inequality, in x, represented on this number line.

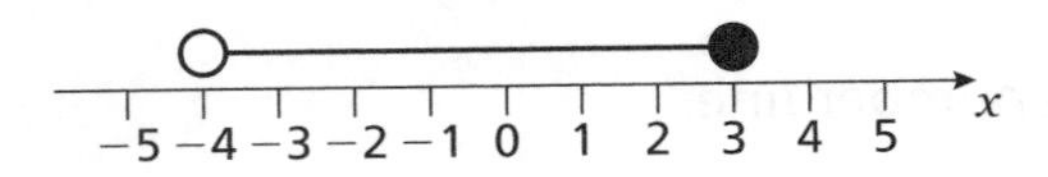

....

b On the number line below, show the inequality $-2 < x \leqslant 4$

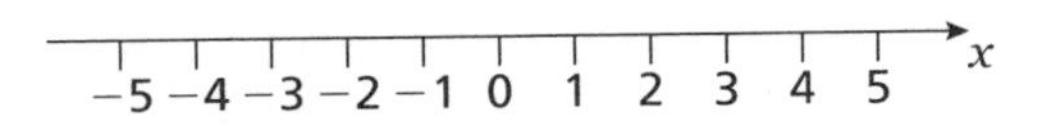

c List the integers that satisfy the inequality $-6 < x \leqslant 2$

....

d Solve the inequality $5x + 8 < 3$

....

4 Solve the inequality $-3x > -9$

....

5 Find the smallest possible integer value of x that satisfies the inequality $2x - 3 > 10$

....

Needs more practice ☐ Almost there ☐ I'm proficient! ☐

6.1 Term-to-term and position-to-term rules

GCSE LINKS
AH: 2.5 Term-to-term and position-to-term definitions; **BH:** Unit 2 7.5 Term-to-term and position-to-term definitions; **16+:** 7.1 Continue and give term-to-term rule for number patterns

By the end of this section you will know how to:

* Find and use the term-to-term rule for a number sequence
* Use a position-to-term rule to find a term in a number sequence

Key points

* A **number sequence** is a pattern of numbers that follows a rule.
* The numbers in a number sequence are called **terms.**
* The **term-to-term rule** for a number sequence says how you can find one term using the term before it.
* The **position-to-term rule** is the rule that connects the value of a term to its position in the number sequence.

Remember this
Number sequences can be continued by adding, subtracting, multiplying and dividing.

Guided

1 Here are the first four terms of a number sequence.

3 8 13 18

a Write down the next two terms in this number sequence.

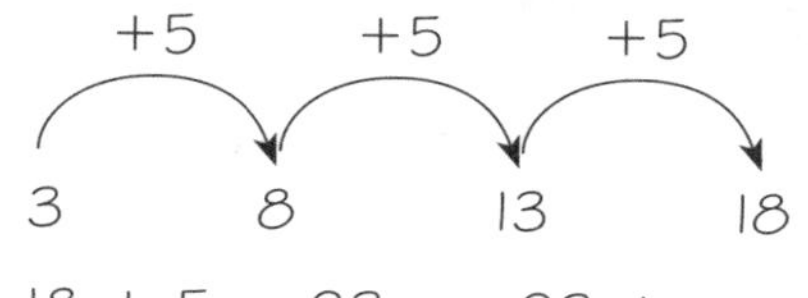

18 + 5 = 23 23 + ... = ... Answer: 23, ...

b What rule do you use to find the next number in the number sequence?

Add ...... to each term

c Find the 10th term in the number sequence.

Method 1

3, 8, 13, 18, 23, 28, 33, 38, 43, **48**

Method 2

Term 1 3

Term **2** 3 + (1 × 5)

Term **3** 3 + (**2** × 5)

...

Term 10 3 + (........ × 5) = ...

Remember this
Method 2 is an example of the position-to-term rule.

Practice

2 Here are the first four terms of a number sequence.

4 8 16 32

a Write down the next two terms in this number sequence.

b What rule do you use to find the next number in the number sequence?

c Find the 7th term in the number sequence.

3 Here are the first four terms of a number sequence.

14 22 30 38

a Write down the next two terms in this number sequence.

b What rule do you use to find the next number in the number sequence?

c Find the 7th term in the number sequence.

Guided

4 Here are the first four terms of a number sequence.

10 100 1000 10 000

Find the 6th term in the number sequence

1st term	2nd term	3rd term	4th term
10	100	1000	10 000
10^1	10^2	10^3	10^4

6th term = 10^6 = ...

Practice

5 Here are the first four terms of a number sequence.

1 8 27 64

Find the 10th term in the number sequence.

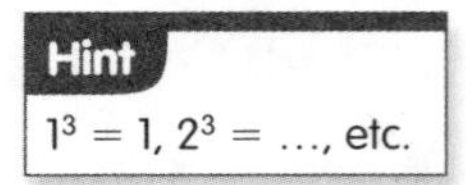

6 Here are the first four terms of a number sequence.

3 9 27 81

Find the 6th term in the number sequence.

7 Here are the first four terms of a number sequence.

1 4 16 64

Find the 6th term in the number sequence.

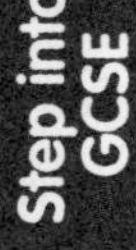

8 Here are the first four terms of a number sequence.

112 108 104 100

a Write down the next term in this number sequence.

....

b Write down the 8th term in this number sequence.

....

c Explain why 7 cannot be a term of this number sequence.

....

....

9 Here are the first four terms of a number sequence.

0 3 8 15

Find the 12th term in the number sequence.

Hint
1st term = $1^2 - 1$,
2nd term = $2^2 - 1$, etc.

....

Needs more practice ☐ Almost there ☐ I'm proficient! ☐

6.2 The nth term of an arithmetic sequence

By the end of this section you will know how to:

* Find and use the nth term of an arithmetic sequence

GCSE LINKS
AH: 2.6 The nth term of an arithmetic sequence; **BH:** Unit 2 7.6 The nth term of an arithmetic sequence; **16+:** 7.2 Find the nth term of a number pattern

Key points

* The **nth term** of an **arithmetic** sequence = (**difference between terms**) × n + **zero term**.

Remember this
An **arithmetic** sequence is one where each successive term is found by adding or subtracting the same fixed number.

Remember this
The **zero term** is the term before the first term. You need to find it in order to find the nth term.

Guided

1 Here are the first four terms of a number sequence.

5 9 13 17

a Write down an expression, in terms of n, for the nth term of the arithmetic sequence.

zero term	1st term	2nd term	3rd term	4th term
1	5	9	13	17

−4 +4 +4 +4

nth term = (difference between terms) × n + zero term

Difference between terms = +4

Zero term = 5 − 4 = 1

So nth term = $4n + 1$

b What is the 50th term in the sequence?

nth term = $4n + 1$

For the 50th term $n = 50$

50th term = 4 × 50 + 1 = ... + 1 = ...

Practice

2 Here are the first four terms of a number sequence.

5 8 11 14

a Write down an expression, in terms of n, for the nth term of the arithmetic sequence.

b What is the 100th term in the sequence?

3 Here are the first four terms of a number sequence.

112 110 108 106

a Write down an expression, in terms of n, for the nth term of the arithmetic sequence.

Remember this

The numbers in the sequence **decrease**, so the difference between the terms is **negative.**

b What is the 50th term in the sequence?

4 Here are the first four terms of a number sequence.

95 90 85 80

a Write down an expression, in terms of n, for the nth term of the arithmetic sequence.

b What is the 20th term in the sequence?

Guided

5 The nth term of a number sequence is $3n - 7$.

a What are the first four terms of the number sequence?

nth term $= 3n - 7$

First term $n = 1$ $3 \times 1 - 7 = 3 - 7 =$ ....

Second term $n = 2$ $3 \times 2 - 7 = 6 - 7 =$ ....

Third term $n = 3$ $3 \times$ .... $- 7 =$ .... $- 7 =$ ....

Fourth term $n =$ .... $=$ .... $=$ ....

Answer: first four terms are: ....

b Which term of the sequence is equal to 65?

$3n - 7 = 65$

$3n - 7 + 7 = 65 + 7$

$3n =$ ....

$n =$ ....

Answer: the .... th term is 65

6 The nth term of a number sequence is $5n + 4$.

a What are the first four terms of the number sequence?

b Which term of the sequence is equal to 59?

7 The nth term of a number sequence is $52 - 4n$.

a What are the first four terms of the number sequence?

b Which term of the sequence is equal to 0?

8 The nth term of a number sequence is $3n^2$.

a What is the 5th term of the sequence?

....

b Is 300 a term of the sequence? Explain your answer.

....

Don't forget!

* Complete the sentences below using the words in the box.

zero term	term-to-term	difference between terms	terms	position-to-term	rule

* A number sequence is a pattern of numbers that follows a ....
* The numbers in a number sequence are called ....
* The .... rule for a number sequence says how you can find one term from the term before it.
* The .... rule is the rule that connects the value of a term to its position in the number sequence.
* The nth term of an arithmetic sequence = (...) $\times n +$ ....
* Match each expression for the nth term of a sequence with the correct first term of that sequence.

$4n + 6$	$3n - 2$	$5 - 2n$	$8 - 3n$
1	5	10	3

Exam-style questions

1 Here are the first four terms of a sequence

6 11 16 21

a Write down the next two terms in this sequence.

....

b Write down an expression, in terms of n, for the nth term of this sequence.

....

c What is the 50th term of this sequence?

....

2 The nth term of a sequence is given by the expression $2n - 5$

a Write down the first two terms of the sequence.

....

b Is 50 a term of the sequence? Explain your answer.

....

....

3 Here are the first four terms of a sequence

$\frac{1}{10}$ $\frac{1}{100}$ $\frac{1}{1000}$ $\frac{1}{10\,000}$

a Write down the next two terms in this sequence.

....

b What is the 8th term of this sequence?

....

4 Here are the first four terms of a sequence

99 91 83 75

a Write down the next two terms in this sequence.

....

b Write down an expression, in terms of n, for the nth term of this sequence.

....

c What is the 20th term of this sequence?

....

5 The nth term of a sequence is given by the expression $10 - 6n$

a Write down the first two terms of the sequence.

....

b Which term of the sequence is equal to -26?

....

6 Here are the first four terms of a sequence

100 81 64 49

a Write down the next two terms in this sequence.

....

b What is the 10th term of this sequence?

....

7 Here are the first four terms of a sequence

12 10 8 6

a Write down an expression, in terms of n, for the nth term of this sequence.

....

b What is the 20th term of this sequence?

....

8 The nth term of a sequence is given by the expression $3n + 4$

a Write down the first two terms of the sequence.

....

b Which term of the sequence is equal to 103?

....

c Is 200 a term of the sequence? Explain your answer.

....

....

Needs more practice ☐ Almost there ☐ I'm proficient! ☐

7.1 Finding the gradient

By the end of this section you will know how to:

* Find the gradient of a straight line graph

GCSE LINKS
AH: 15.3 The gradient and y-intercept of a straight line; **BH:** Unit 2 9.3 The gradient and y-intercept of a straight line

Key points

* The **gradient** of a straight line is a measure of its slope.
* The **steeper** the line the **greater** the gradient.
* The gradient of a straight line can be:

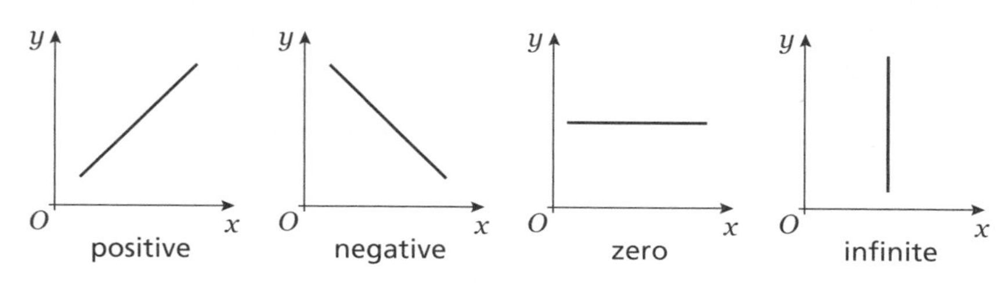

* Gradient of a line $= \dfrac{\text{change in } y \text{ value}}{\text{change in } x \text{ value}}$
* If the line slopes downwards from left to right, the change in the y-value is negative.

You should know

Coordinates are given as (x, y) in alphabetical order. For example:

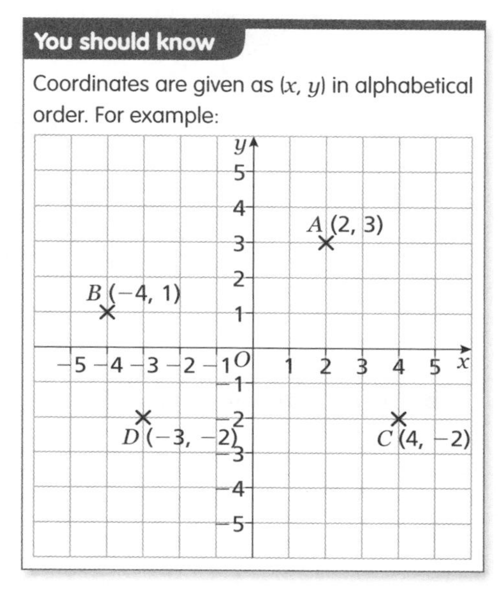

Guided

1 Find the gradient of the straight line joining these points.

a A (2, 1) and B (3, 3)

Remember this
Draw a diagram to help you.

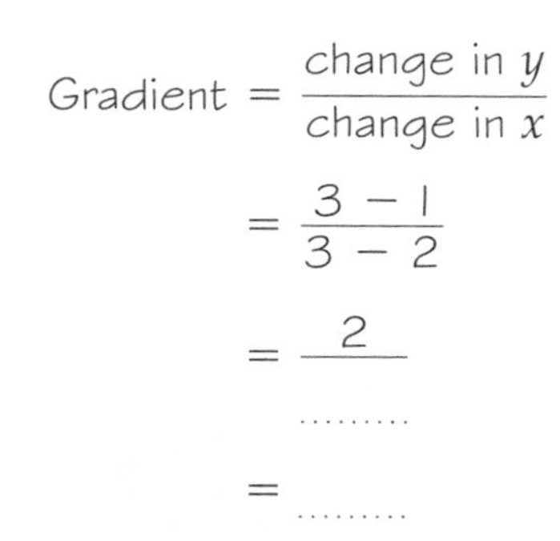

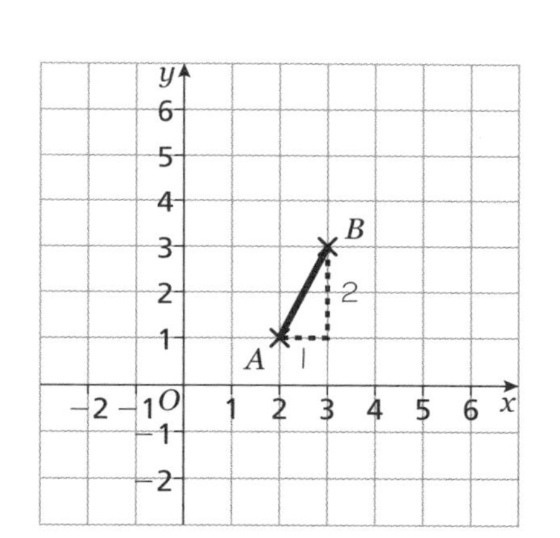

b A (−2, 2) and B (4, 5)

Gradient $= \dfrac{\text{change in } y}{\text{change in } x}$

$= \dfrac{5 - 2}{4 - (-2)}$

$= \dfrac{3}{\dots\dots\dots}$

$= \dfrac{\dots\dots\dots}{\dots\dots\dots}$

Remember this
Check the scale used on each axis when finding the lengths of the sides of the triangle.

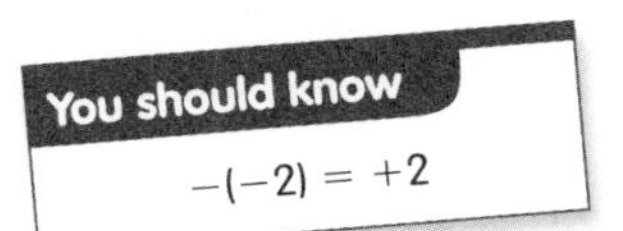

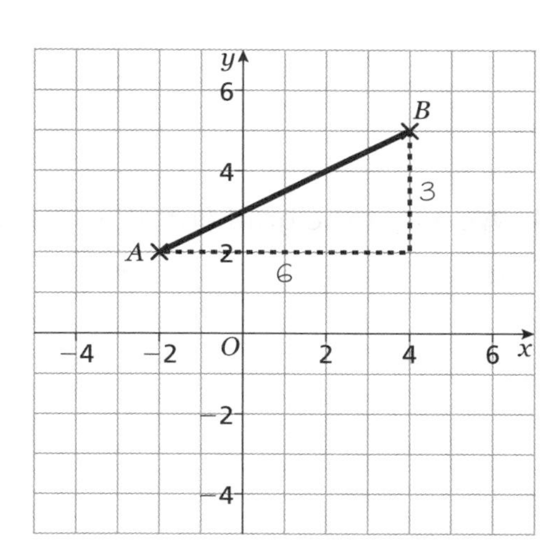

2 Find the gradient of the straight line joining these points.

a $A\ (-3, 6)$ and $B\ (2, -4)$

$$\text{Gradient} = \frac{\text{change in } y}{\text{change in } x}$$

$$= \frac{-4 - 6}{2 - (-3)}$$

$$= -\frac{\ldots\ldots}{\ldots\ldots}$$

$$= -\ldots\ldots$$

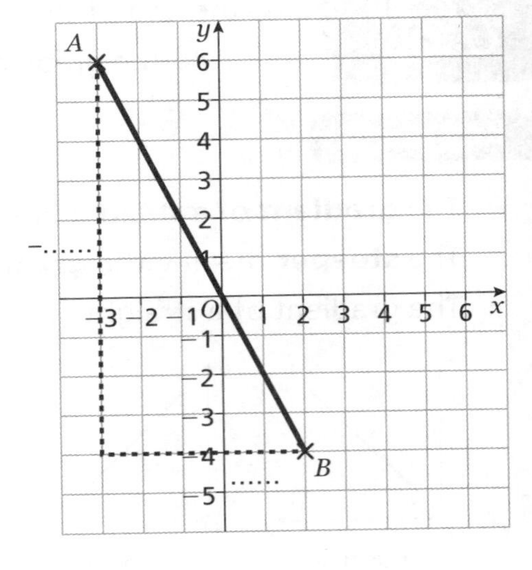

Remember this

As the line slopes downwards from left to right, the change in y is negative; y goes from a larger value to a smaller value.

b $A\ (4, -3)$ and $B\ (6, -5)$

$$\text{Gradient} = \frac{\text{change in } y}{\text{change in } x}$$

$$= \frac{\ldots\ldots}{\ldots\ldots}$$

$$= \frac{\ldots\ldots}{\ldots\ldots}$$

$$= -\ldots\ldots$$

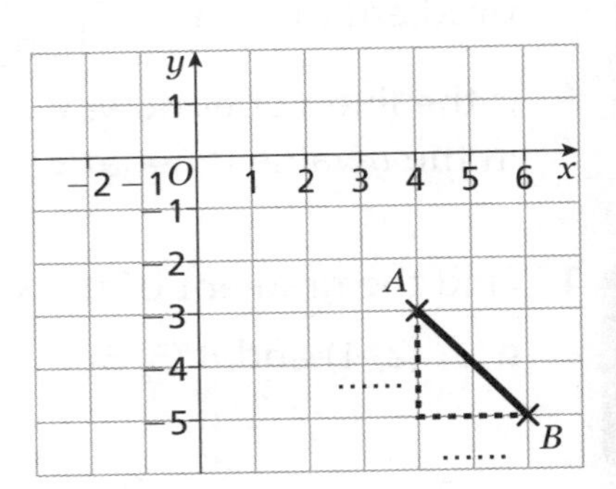

3 Work out the gradient of each line.

a

b

c

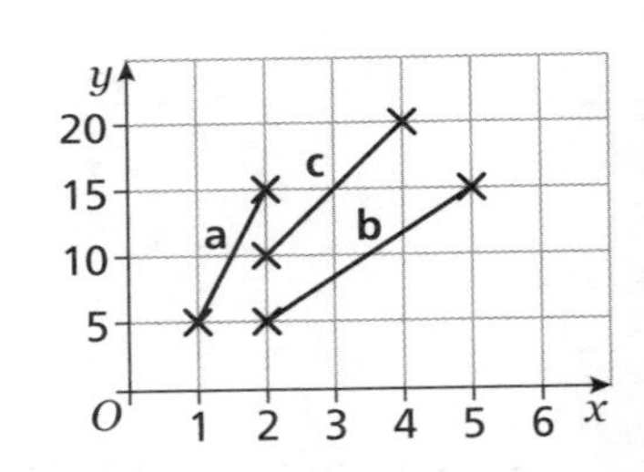

4 Work out the gradient of each line.

a

b

c

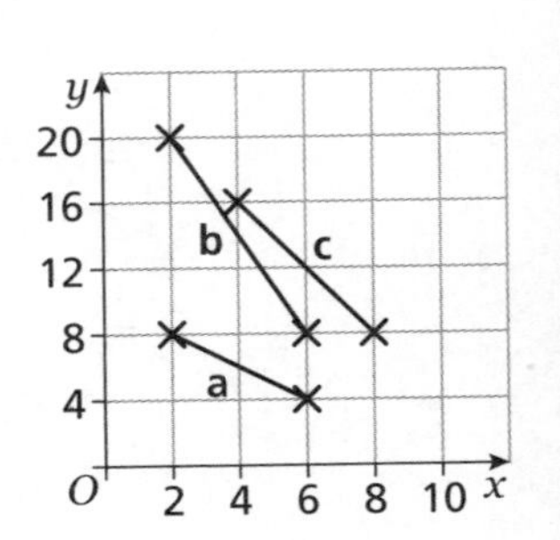

Guided

5 Find the gradient of this straight line graph.

Gradient = $\frac{\text{change in } y}{\text{change in } x}$

$= \frac{16 - 4}{6 - 2}$

$= \frac{12}{\dots\dots\dots}$

$= \dots\dots\dots$

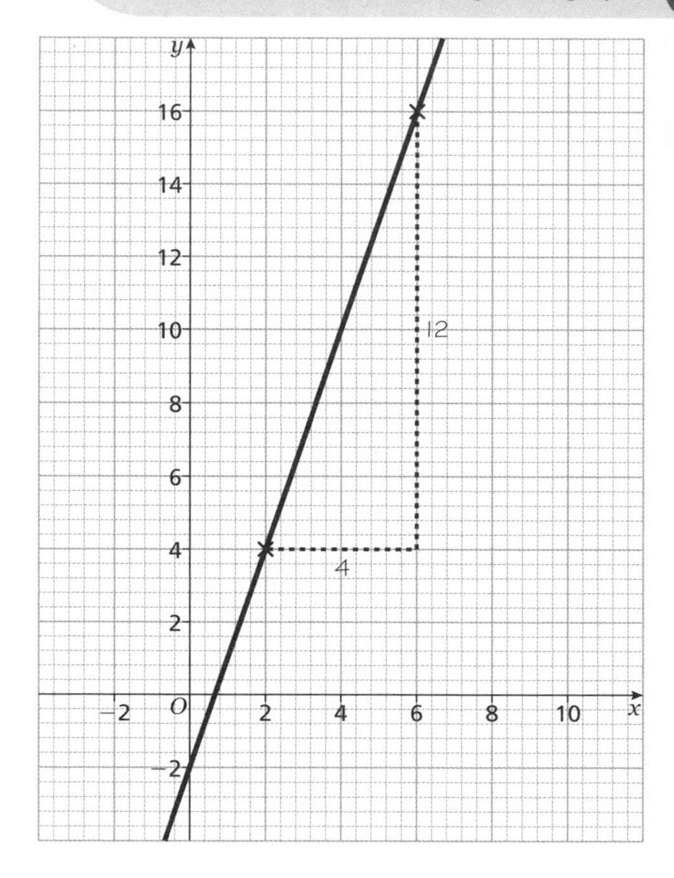

Remember this

Draw any right-angled triangle with a base which is a whole number of units.

6 Find the gradient of this straight line graph.

Gradient = $\frac{\text{change in } y}{\text{change in } x}$

$= \frac{-2 - 6}{2 - (-2)}$

$= -\frac{\dots\dots\dots}{\dots\dots\dots}$

$= \dots\dots\dots$

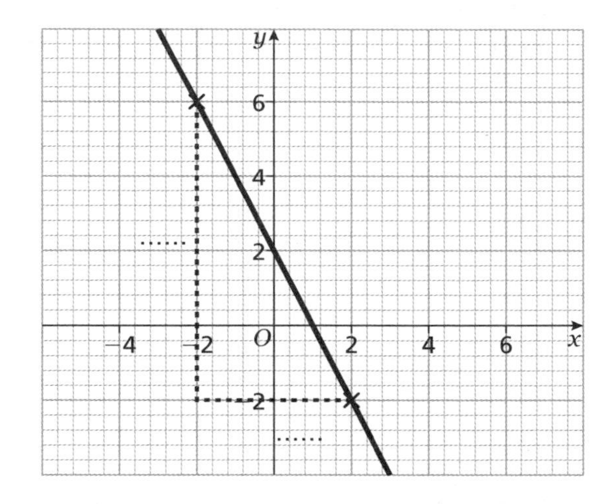

Remember this

Wherever you draw the right-angled triangle on the straight line you will always get the same answer for the gradient.

Remember this

In Q5 the gradient is 3 or $\frac{3}{1}$, which means for every 1 you go to the right you go **up** 3. In Q6 the gradient is −2 or $-\frac{2}{1}$ which means for every 1 you go to the right you go **down** 2.

Practice

7 Find the gradient of these straight line graphs

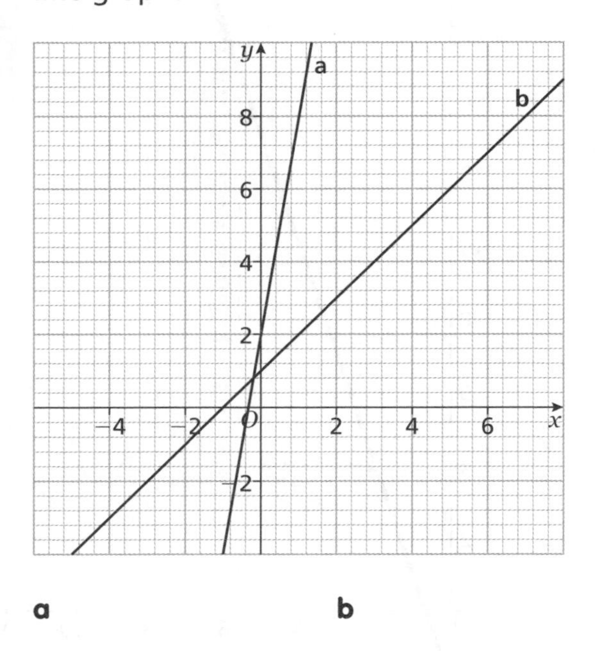

a

b

8 Find the gradient of these straight line graphs.

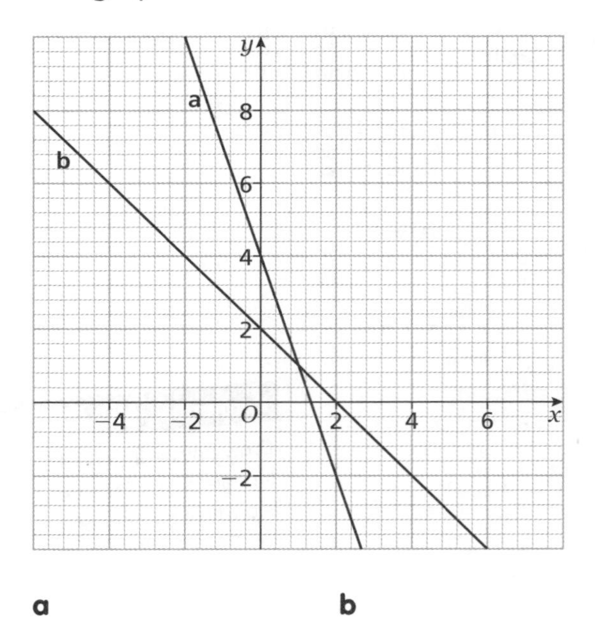

a

b

9 A has coordinates (0, 2). B has coordinates (−2, 6).
Work out the gradient of the line that passes through A and B.

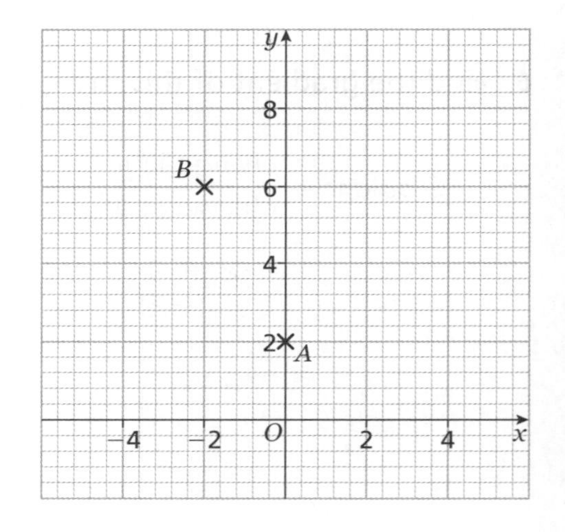

Needs more practice ☐ Almost there ☐ I'm proficient! ☐

7.2 Interpreting the gradient

By the end of this section you will know how to:

* Interpret the gradient of a real-life graph

GCSE LINKS

AH: 15.3 The gradient and y-intercept of a straight line; **BH:** Unit 2 9.3 The gradient and y-intercept of a straight line; **16+:** 8.4 Interpret and draw graphs you meet in everyday life

Key points

* A real-life graph has units on both axes.
* The units for the gradient are the units on the vertical axis, divided by the units on the horizontal axis.

Guided

1 This is an approximate conversion graph to convert between British pounds (£) and euros (€).

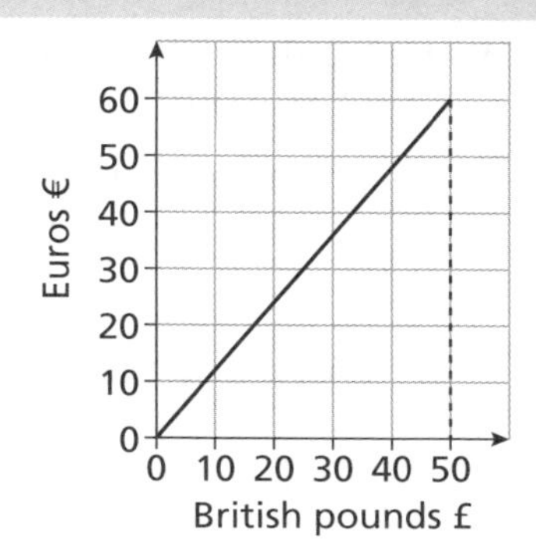

a Work out the gradient of the graph.

$$\text{Gradient} = \frac{60 - 0}{50 - 0}$$

$$= \frac{60}{50}$$

$$= \ldots\ldots$$

Remember this

You can use the whole graph to draw a triangle.

b What does the gradient of the graph represent?

$$\text{Gradient} = \frac{\text{Euros}}{\text{British pounds}}$$

Units are euros per pound, so this gives the conversion rate £1 = ... euros

2 The graph shows the charges for a car hire company. The company charges a basic fee plus a cost per day.

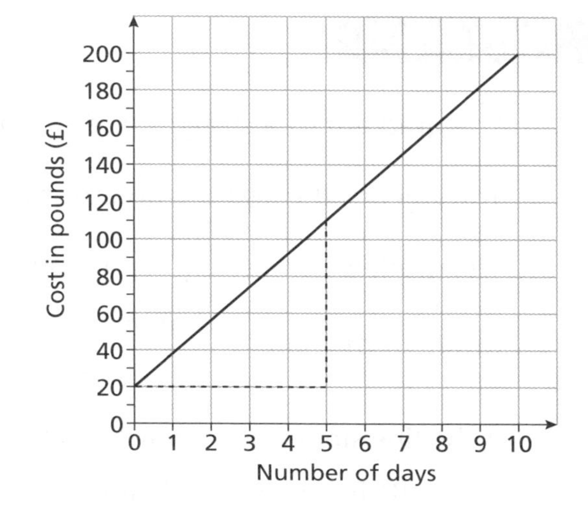

a What is the basic fee?

Basic fee = cost when number of days = 0

So basic fee = £.........

b Work out the gradient of the line.

$$\text{Gradient} = \frac{110 - 20}{5 - 0}$$

$$= \frac{\ldots\ldots}{\ldots\ldots}$$

$$= \ldots\ldots$$

c What does the gradient of the line represent?

$$\text{Gradient} = \frac{\text{cost}}{\text{number of days}}$$

Units are cost per ...., so cost per .... = ...

Practice

3 This is a conversion graph to convert between pounds and kilograms.

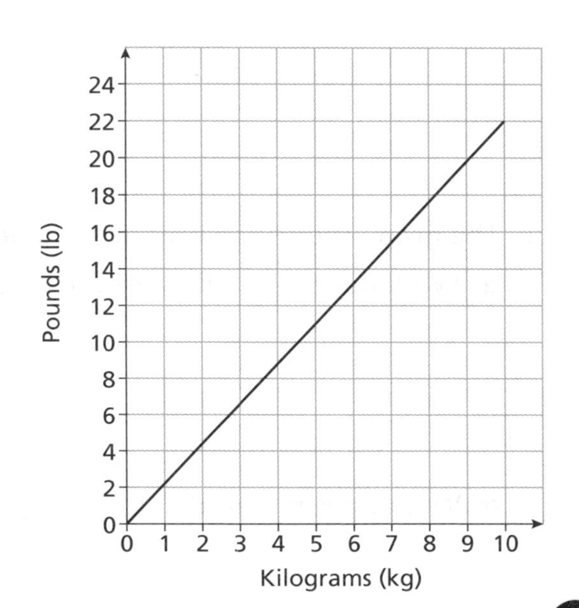

a Work out the gradient of the graph.

b What does the gradient of the graph represent?

4 The graph shows the charges for a taxi company.
The taxi company charges a basic cost plus a cost per mile.

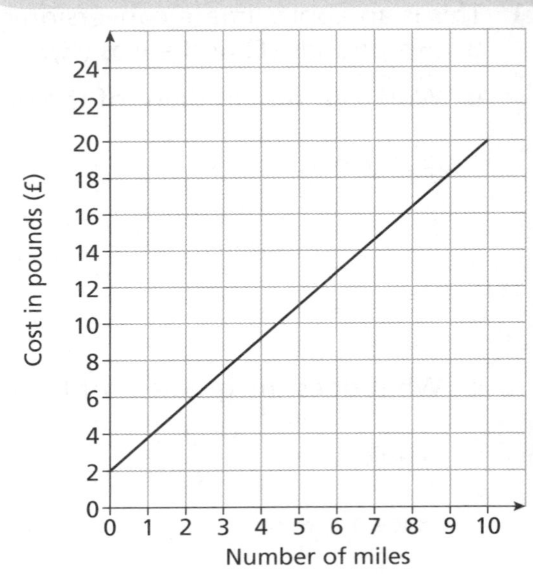

a What is the basic cost?

b Work out the gradient of the line.

c What does the gradient of the line represent?

Don't forget!

* The gradient of a straight line is a measure of its ....
* The steeper the line the greater the ....
* The gradient of a straight line can be ...., ..., ... or ....
* Gradient of a line $= \dfrac{\text{change in } \ldots\ldots\ldots\ldots}{\text{change in } \ldots\ldots\ldots\ldots}$
* If the line slopes downwards from left to right, the change in the y-value is ....
* The units for the gradient are the units on the .... axis, divided by the units on the .... axis.

Exam-style questions

1 Find the gradient of the straight line joining the points A $(-1, 2)$ and B $(3, 8)$.

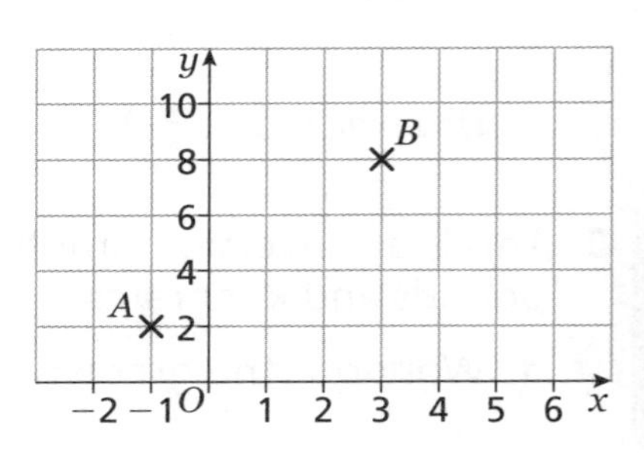

....

2 The graph shows the distance in miles travelled by a car over a period of time in hours.

a Work out the gradient of the graph.

Distance (miles)
120
100
80
60
40
20
0
0 1 2 3 4
Time (hours)

....

b What does the gradient of the graph represent?

....

3 Find the gradient of this straight line graph.

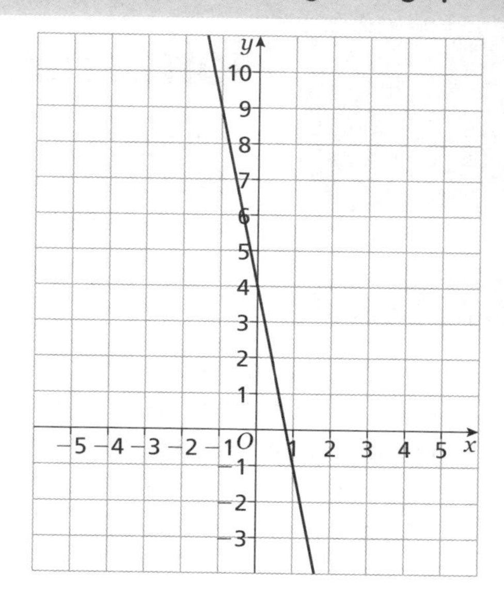

....

4 Find the gradient of the straight line joining the points A (−2, 8) and B (4, 2).

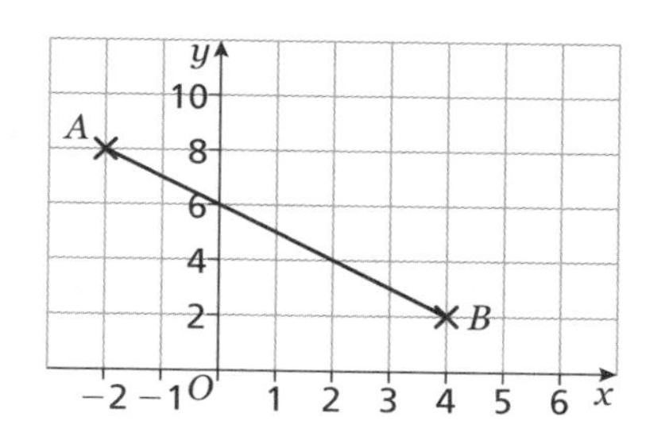

....

5 The graph shows the amount of money students are paid for working in a local shop. They are paid a standard wage for working 4 hours, plus a fee for every additional hour they work more than 4 hours.

a What is the standard wage?

£ ....

b Work out the gradient of the graph.

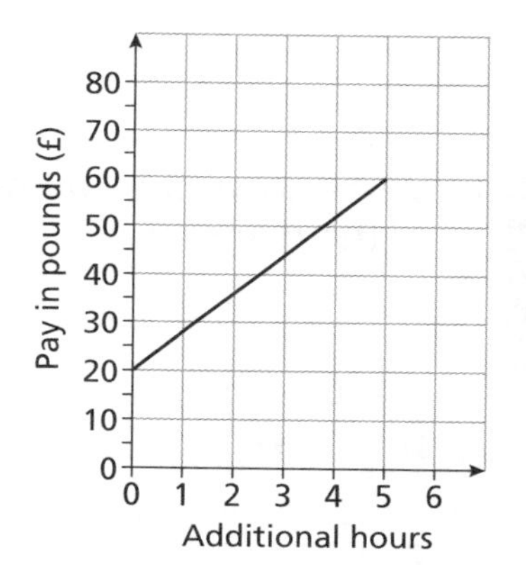

....

c What does the gradient of the graph represent?

....

6 Find the gradient of the straight line joining the points A (−4, −2) and B (4, 8).

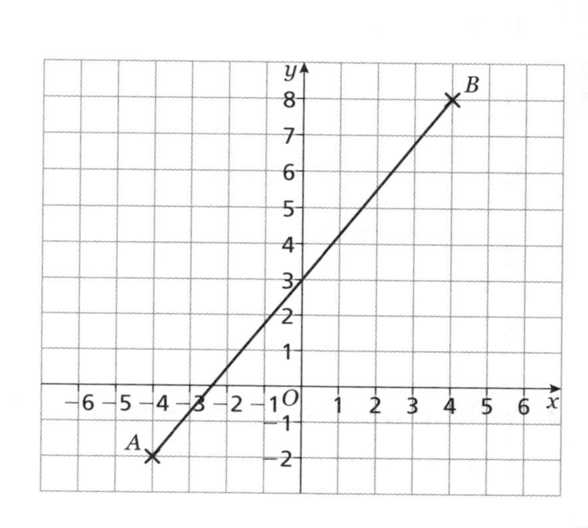

....

8.1 Horizontal and vertical lines

By the end of this section you will know how to:

* Recognise, plot and draw graphs of the form $x = a$ and $y = b$

GCSE LINKS

AH: 15.1 Drawing straight line graphs by plotting points, 15.3 The gradient and y-intercept of a straight line, 15.4 The equation $y = mx + c$; **BH:** Unit 2 9.1 Drawing straight line graphs by plotting points, 9.3 The gradient and y-intercept of a straight line, 9.4 The equation $y = mx + c$; **16+:** 8.1 Draw straight line graphs

Key points

* A vertical line on the grid has an equation of the form $x = a$.
  For example, the equation of this line is $x = 3$ as it passes through the point where $x = 3$ on the x-axis.

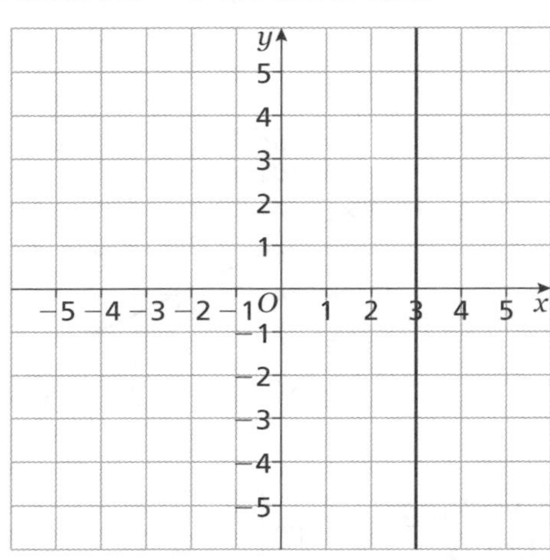

Remember this

On a vertical line, for every point on the line, the x-coordinate is always the same; similarly the y-coordinate of a horizontal line is always the same.

* A horizontal line on the grid has an equation of the form $y = b$.
  For example, the equation of this line is $y = -2$ as it passes through the point where $y = -2$ on the y-axis.

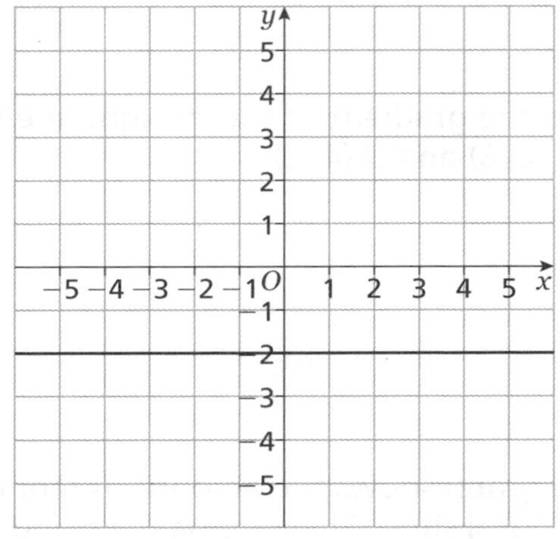

Guided

1 Write down the equation of each of the lines **a** to **d** in the diagram.

a $y = 3$

b $y = -$ ....

c $x =$ ....

d ....

Practice

2 Write down the equation of each of the lines **a** to **d** in the diagram.

a 

b ....

c ....

d ....

3 Write down the equation of

a the x-axis

b the y-axis

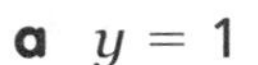

4 On the grid draw and label the lines with these equations.

a $y = 1$ **b** $y = -1$

c $x = 3$ **d** $x = -3$

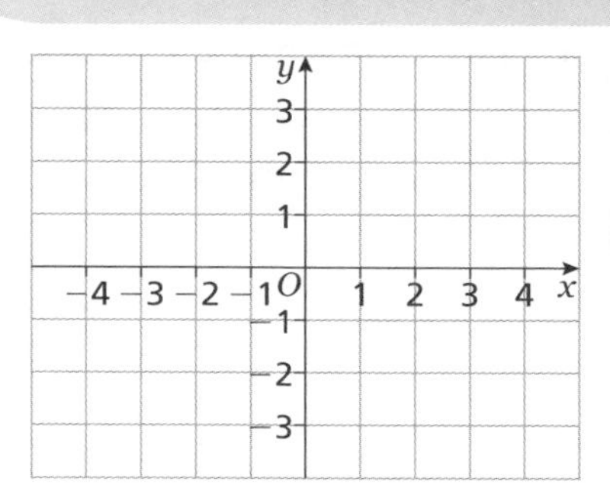

Needs more practice ☐ Almost there ☐ I'm proficient! ☐

8.2 The equation $y = mx + c$

By the end of this section you will know how to:

* Recognise graphs of the form $y = mx + c$

GCSE LINKS

AH: 15.1 Drawing straight line graphs by plotting points, 15.3 The gradient and y-intercept of a straight line, 15.4 The equation $y = mx + c$; **BH:** Unit 2 9.1 Drawing straight line graphs by plotting points, 9.3 The gradient and y-intercept of a straight line, 9.4 The equation $y = mx + c$; **16+:** 8.1 Draw straight line graphs

Key points

* The equation of a straight line can be written as $\boldsymbol{y = mx + c}$.
* The gradient of the straight line is m.
* The straight line crosses the y-axis at $(0, c)$.
* The point where the straight line crosses the y-axis is called the **y-intercept**.

Remember this

In the equation $y = mx + c$, the number in front of the x is the gradient, and the number on its own is the y-intercept.

Guided

1 Write down the gradient of the lines with these equations.

a $y = 3x$

$y = 3x$

gradient $= 3$

Hint

The gradient of a straight line has a **constant** value, i.e. it is the same at every point along the line. There should be no x in your answer.

b $y = -4x$

$y = -4x$

gradient $= -$ ....

c $y = x$

$y = x$

gradient $=$ ....

You should know

$y = x$ is the same as $y = 1x$.

2 Write down the gradient and the y-intercept of the lines with these equations.

a $y = 2x - 3$

$m =$ gradient $= 2$

$c = y$-intercept $= -3$

b $y = 5 - 3x$

$m =$ gradient $= -$ ....

$c = y$-intercept $=$ ....

c $y = 7x$

$m =$ gradient $=$ ....

$c = y$-intercept $=$ ....

Practice

3 Write down the gradient and the y-intercept of the lines with these equations.

a $y = 8x + 7$ **b** $y = x - 9$ **c** $y = 4 - 6x$

Guided

4 Write down the equation of each of these straight lines.

a gradient $= 4$ and y-intercept $= -3$ $m = 4$ and $c = -3$; so equation is $y = 4x - 3$

b gradient $= -2$ and y-intercept $= 5$ $m = -2$ and $c =$ ....; so equation is $y = -2x$ ....

c gradient $= 1$ and y-intercept $= -7$ $m =$ .... and $c =$ ....; so equation is $y =$ ....

Practice

5 Write down the equation of each of these straight lines.

a gradient = −5 and y-intercept = 1

b gradient = 7 and y-intercept = −2

c gradient = −4 and y-intercept = 0

Needs more practice ☐ Almost there ☐ I'm proficient! ☐

8.3 Plotting and drawing graphs

By the end of this section you will know how to:

* Draw graphs of the form $y = mx + c$

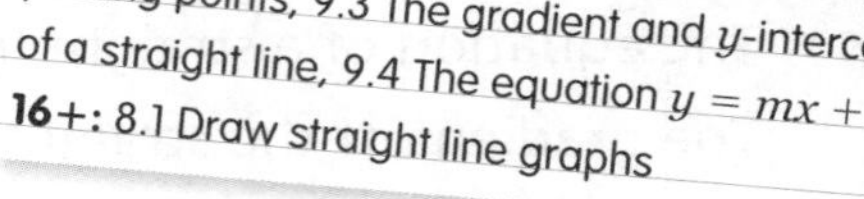

GCSE LINKS

AH: 15.1 Drawing straight line graphs by plotting points, 15.3 The gradient and y-intercept of a straight line, 15.4 The equation $y = mx + c$; **BH:** Unit 2 9.1 Drawing straight line graphs by plotting points, 9.3 The gradient and y-intercept of a straight line, 9.4 The equation $y = mx + c$; **16+:** 8.1 Draw straight line graphs

Key points

* Straight line graphs can be drawn using a table of values.
* Straight line graphs can be drawn using the y-intercept and the gradient.

Remember this

You need to complete or set up a table of values and find the value of y by substituting different values of x into the equation.

Guided

1 a Complete the table of values for $y = 2x + 1$.

x	−1	0	1	2	3	4
$y = 2x + 1$	−1	....	....	5	....	....

Remember this

In a table of values, for a straight line graph, if the increases in the values of x are equal, then the gap between the y values will be the same.

$x = -1$: $y = 2 \times -1 + 1 = -2 + 1 = -1$

$x = 0$: $y = 2 \times 0 + 1 =$ .... $+ 1 =$ ....

$x = 1$: $y = 2 \times 1 + 1 =$ .... + ... = ...

$x = 2$: $y = 2 \times$ .... $+ 1 =$ .... $+ 1 =$ ....

$x = 3$: $y = 2 \times$ .... + ... = ... + ... = ...

$x = 4$: $y =$ .... = ... = ...

b On the grid draw the line with equation $y = 2x + 1$ for values of x from −1 to 4.

You should know

Join the plotted points using a straight line.

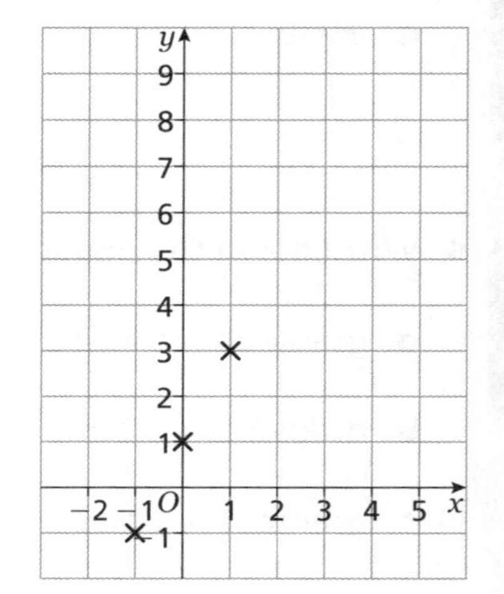

2 On the grid draw the graph of $x + y = 5$, from $x = 0$ to $x = 5$.

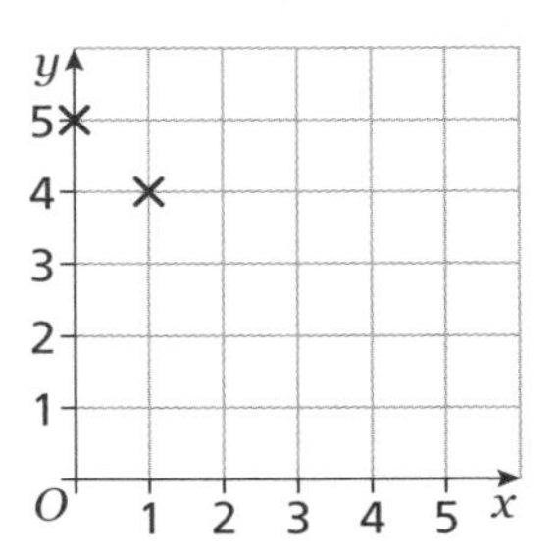

x	0	1	2	3	4	5
y	5	4				

Remember this

Three points are enough to fix a line, so in questions where you have to set up your own table, you could just substitute three values of x between 0 and 5 and find the corresponding values of y to plot.

If $x + y = 5$ then $y = 5 - x$

$x = 0$: $y = 5 - 0 = 5$

$x = 1$: $y = 5 - 1 = 4$

$x = 2$: $y = 5 -$ ……… $=$ ………

$x = 3$: $y = 5 -$ ……… $=$ ………

$x = 4$: $y =$ ………………………… $=$ ………

$x = 5$: $y =$ ………………………… $=$ ………

Hint

First rearrange the equation to make y the subject. Set up a table with values of x between 0 and 5, then substitute these into the equation to find the values of y.

3 a Complete the table of values for $y = 3x - 5$.

x	−1	0	1	2	3
$y = 3x - 5$					

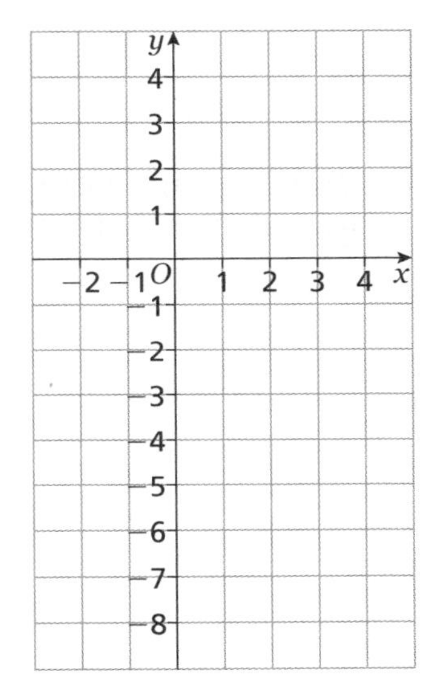

b On the grid draw the line with equation $y = 3x - 5$ for values of x from −1 to 3.

4 a Complete the table of values for $y = -4x + 2$.

x	−1	0	1	2	3
$y = -4x + 2$					

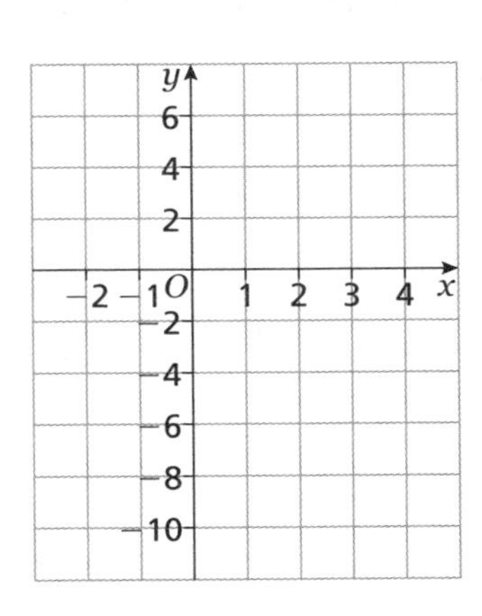

b On the grid draw the line with equation $y = -4x + 2$ for values of x from −1 to 3.

Practice

5 On the grid draw the graph of $x + y = 10$, from $x = 0$ to $x = 10$.

Hint
You could, for example, use values of x in steps of 2 in your table.

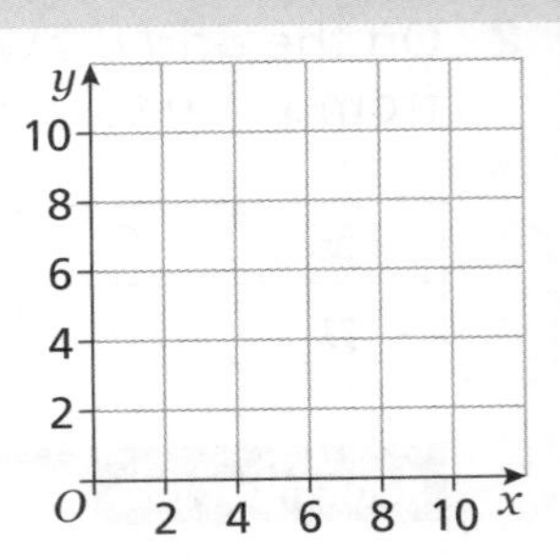

Guided

6 Draw a graph with equation $y = 2x + 4$ for values of x from -1 to 4.

gradient $= m = 2$,

so for every 1 unit you go to the right on the x-axis you go up 2 units on the y-axis

y-intercept $= c = 4$,

so graph crosses 4 on y-axis

Remember this
Make sure you take the scale used on each axis into account when using this method.

Remember this
Draw a straight line through the plotted points.

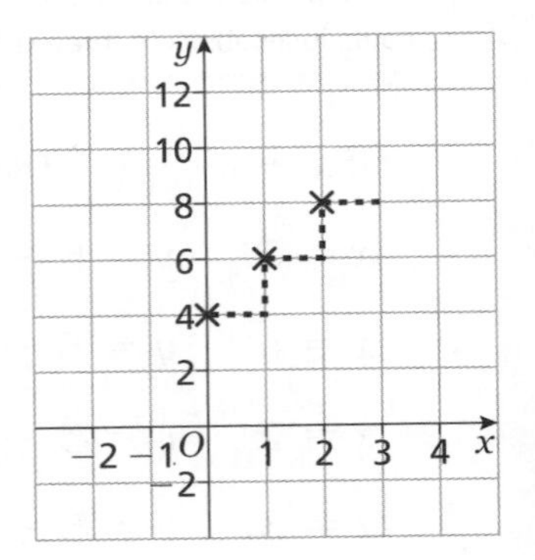

Practice

7 Draw a graph with equation $y = -5x + 5$, for values of x from -1 to 3.

Remember this
Because the gradient is -5, that is, it is negative, to find each successive point from left to right go across 1 unit and **down** 5 units.

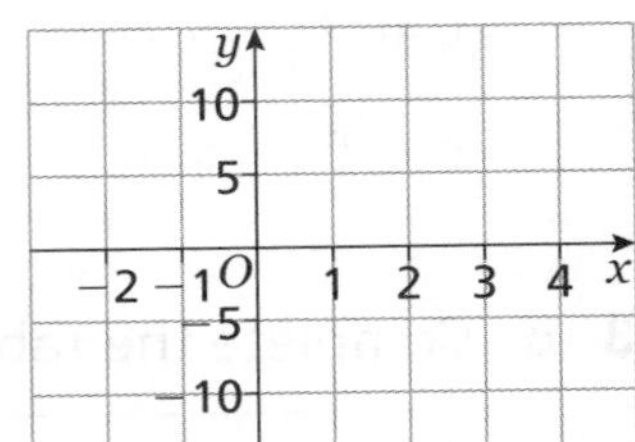

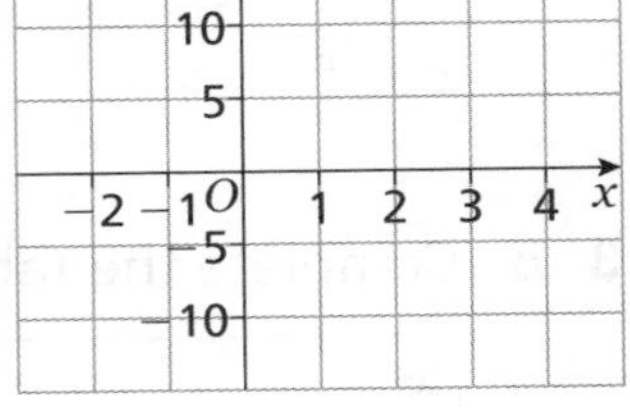

8 Draw a graph with equation $y = 4x - 2$, for values of x from from -1 to 3.

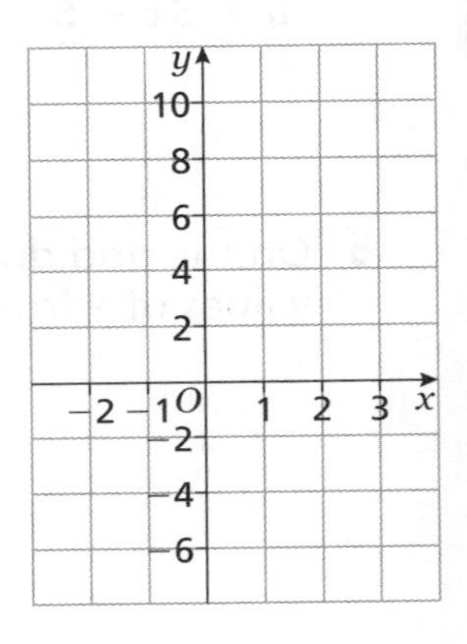

Step into GCSE

9 Draw the graph of $y = -4x - 2$ for values of x from -1 to 3.

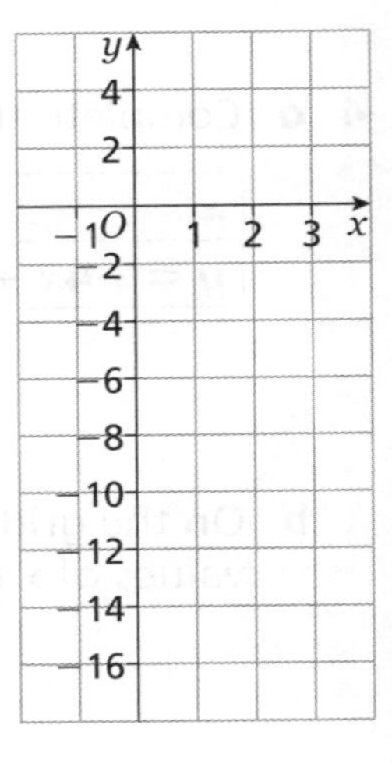

Needs more practice ☐ Almost there ☐ I'm proficient! ☐

8.4 Finding the equation of a straight line graph

By the end of this section you will know how to:

* Find the equation of a given straight line graph

GCSE LINKS
AH: 15.3 The gradient and y-intercept of a straight line, 15.4 The equation $y = mx + c$; **BH:** Unit 2 9.3 The gradient and y-intercept of a straight line, 9.4 The equation $y = mx + c$; **16+:** 8.2 Draw and write the equations of straight line graphs

Key points

* The equation of a straight line is $y = mx + c$.
* You can read off the value of c (the y-intercept) from the straight line graph.
* You can calculate m (the gradient) by drawing in a suitable triangle and using $m = \dfrac{\text{change in } y}{\text{change in } x}$

Guided

1 Write down the equation of each of the lines **a** to **d** in the diagram.

a $c = -5$,

$m = \dfrac{-1 - (-5)}{2 - 0} = \dfrac{4}{2} = 2$

Equation is $y = 2x - 5$

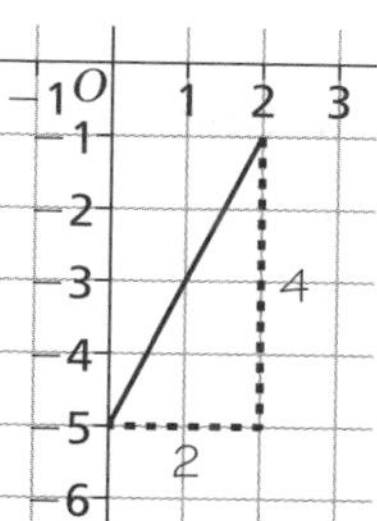

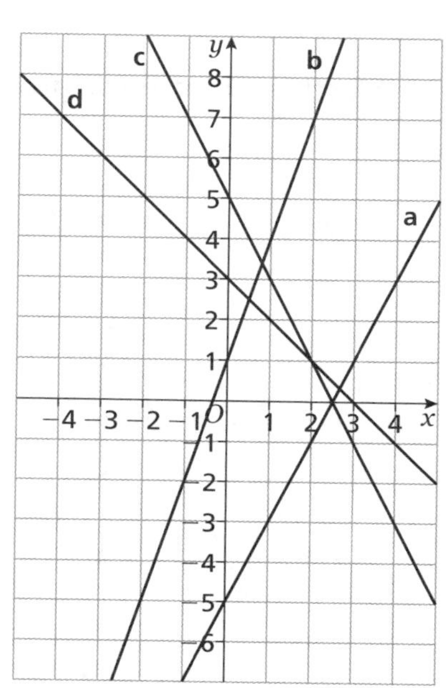

b $c = 1$

$m = \dfrac{7 - 1}{2 - 0} = \dfrac{\ldots\ldots}{\ldots\ldots} = \ldots\ldots$

Equation is $y = \ldots\ldots + 1$

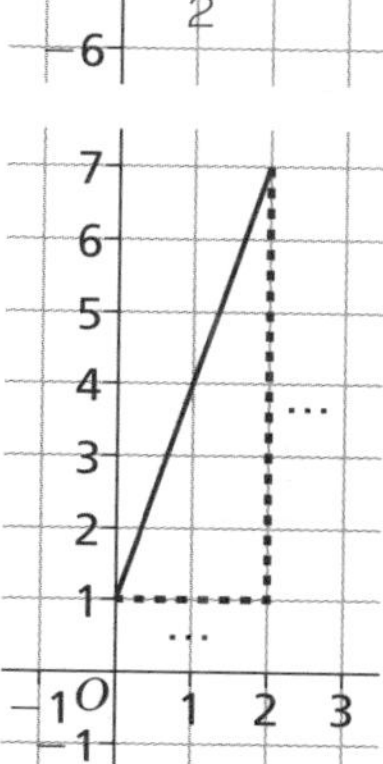

c $c = \ldots\ldots$ $m = \dfrac{\ldots\ldots}{\ldots\ldots} = \ldots\ldots$ Equation is $y = \ldots\ldots\ldots\ldots$

d $c = \ldots\ldots$ $m = \dfrac{\ldots\ldots}{\ldots\ldots} = \ldots\ldots$ Equation is $y = \ldots\ldots\ldots\ldots$

Practice

2 Write down the equation for each of the lines **a** to **d** in the diagram.

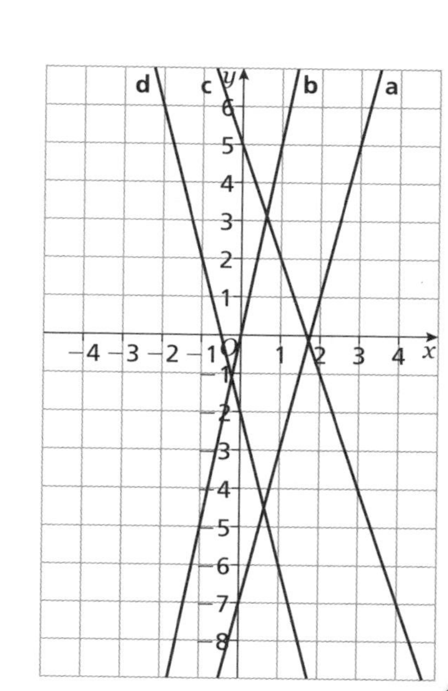

a

b

c

d

Don't forget!

* A vertical line has an equation of the form ....
* A horizontal line has an equation of the form ....
* The x-axis has equation ....
* The y-axis has equation ....
* The equation of a straight line can be written as $y = mx + c$ where the gradient of the straight line is .... and the y-intercept is ....
* The gradient of a straight line graph can be calculated using

  gradient $= \dfrac{\text{..............................}}{\text{..............................}}$

* Match each of these graphs to the correct equation.

A

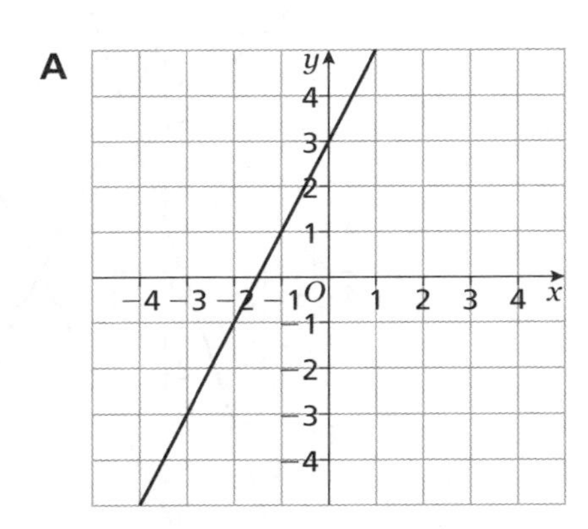

$y = -2x + 3$

B

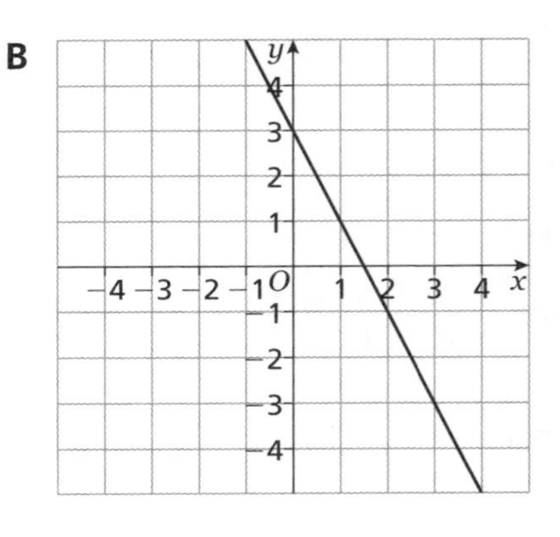

$y = 4x + 1$

C

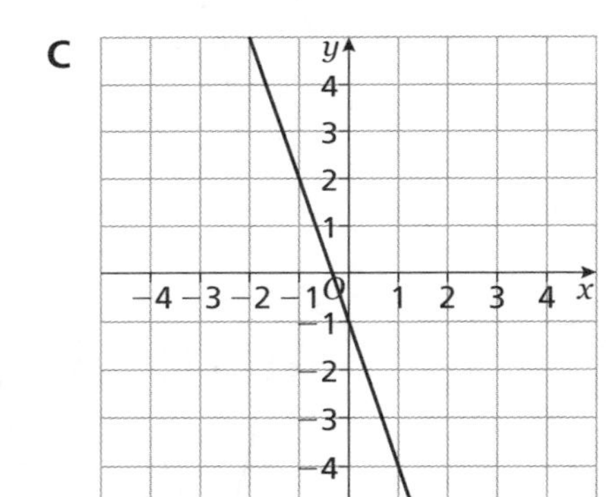

$y = 2x + 3$

D

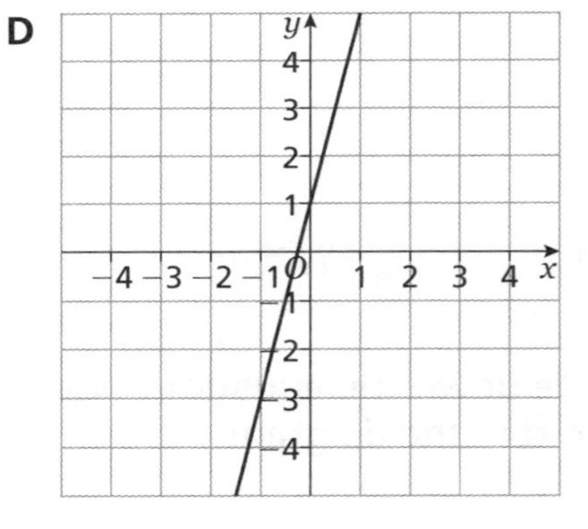

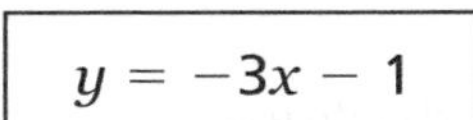

Exam-style questions

1 On the grid draw the line with equation $y = 3x - 4$ for values of x from -1 to 3

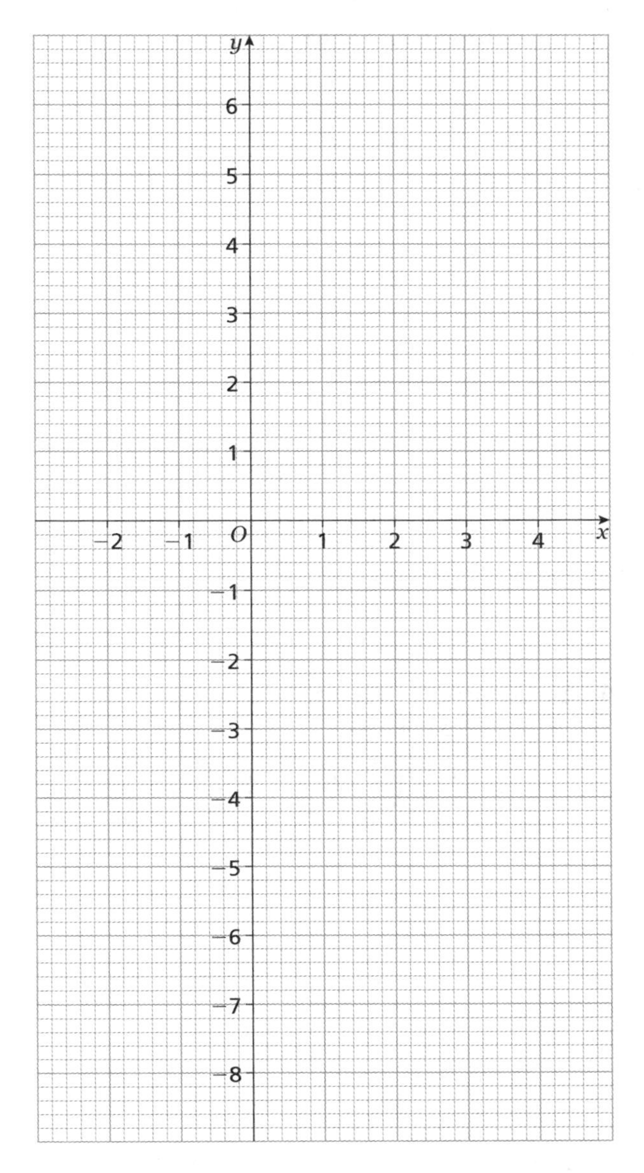

2 The line **L** is shown on the grid.

a Work out the gradient of the line **L**.

....

b Find an equation of the line **L**.

....

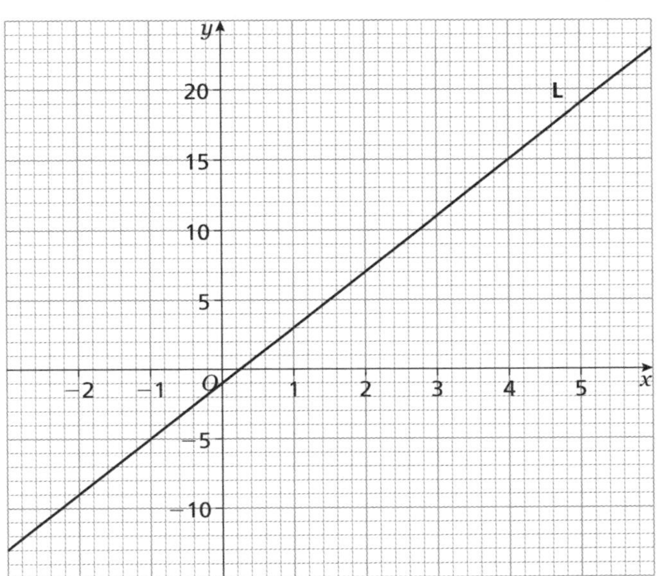

3 On the grid draw the line with gradient -2, which passes through the point (0, 4) for values of x from -2 to 3

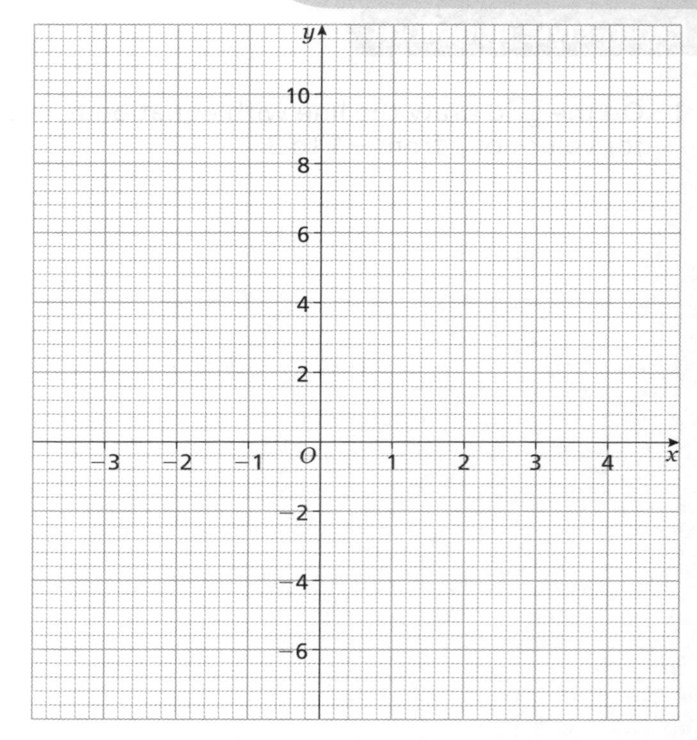

4 a Complete the table of values for $y = -3x + 7$

x	-1	0	1	2	3	4
y						

b On the grid draw the line with equation $y = -3x + 7$ for values of x from -1 to 4

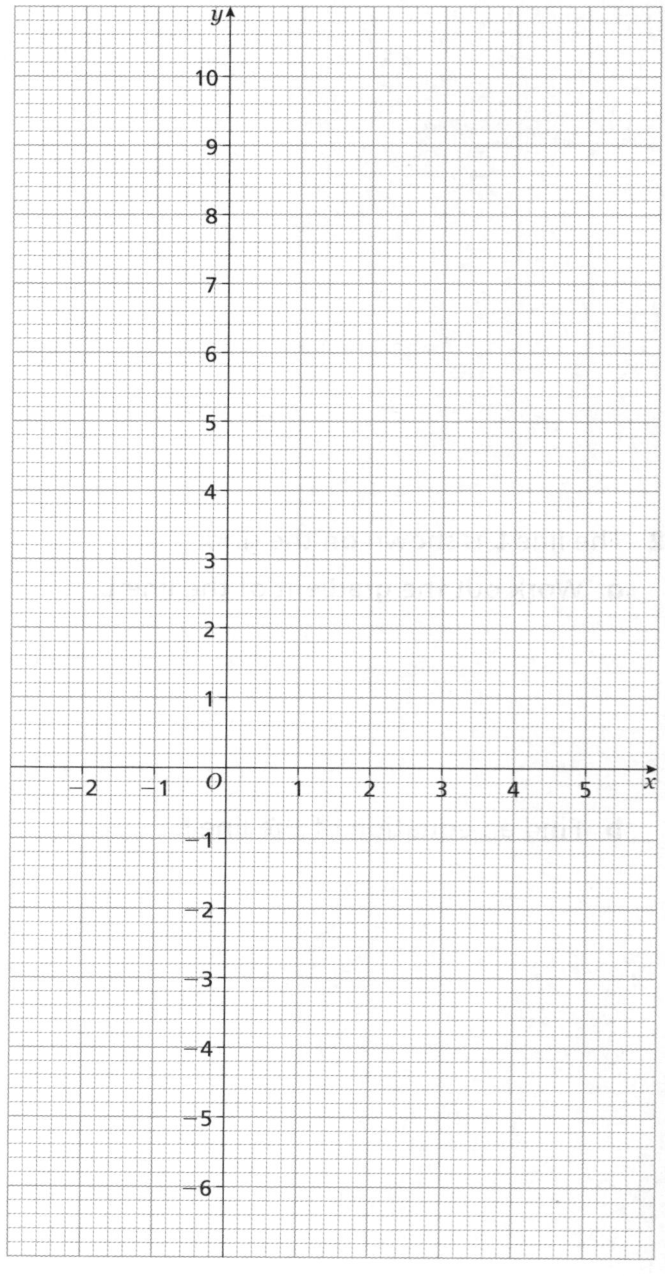

5 On the grid draw the line with equation $y = -2x + 2$ for values of x from -2 to 3

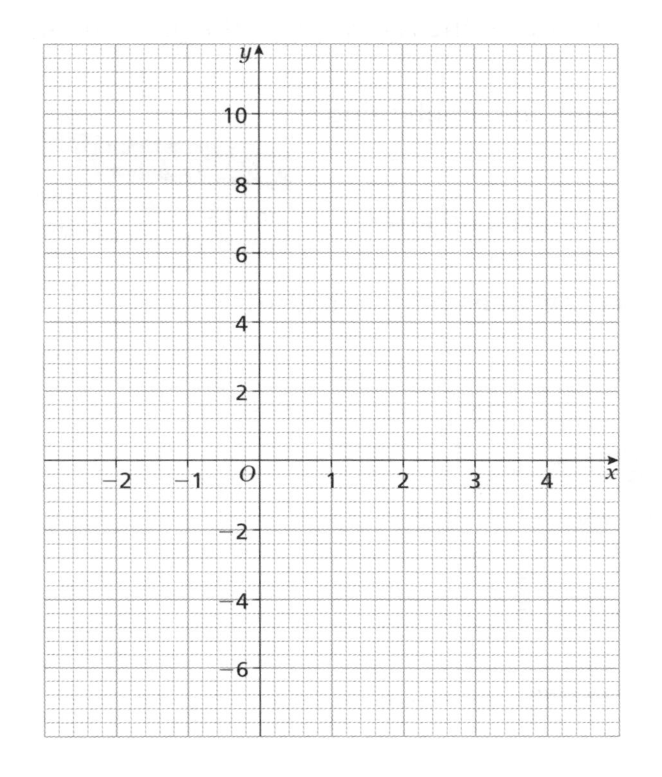

6 On the grid draw the line with equation $x + y = 8$ for values of x from 0 to 8

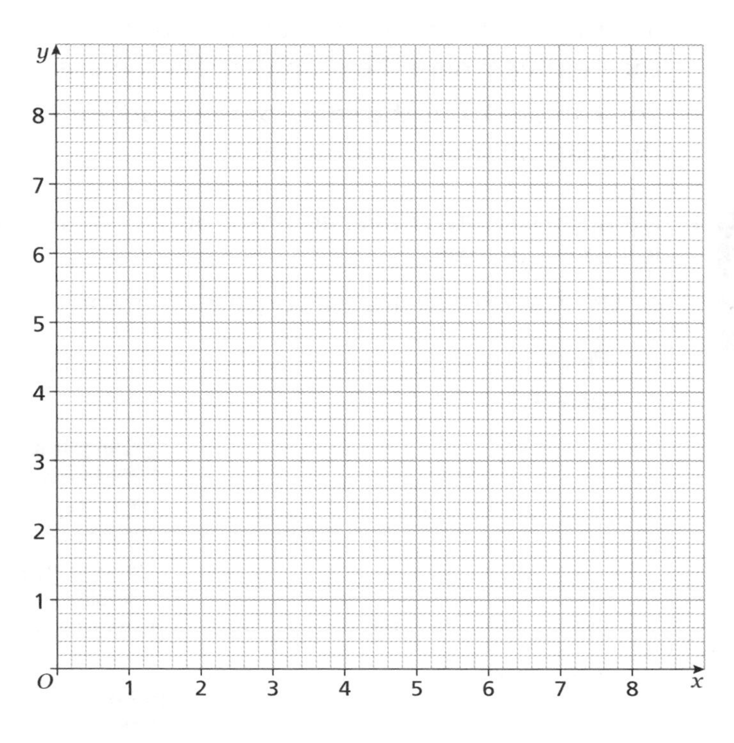

7 On the grid draw the line with gradient 5, which passes through the point (0, 3) for values of x from -1 to 3

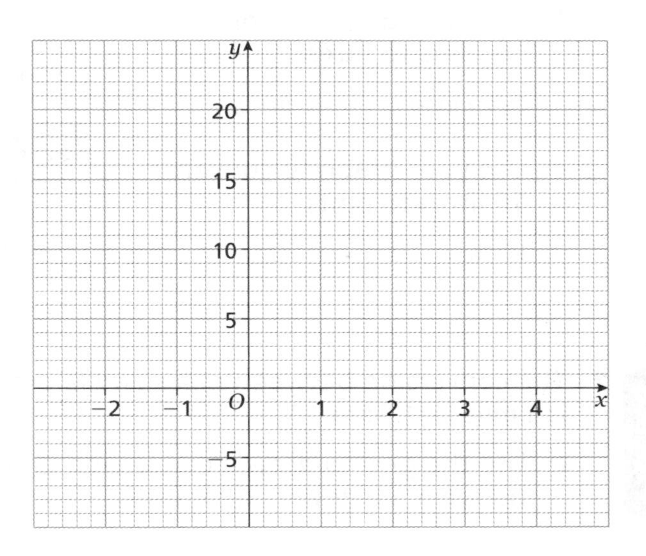

9.1 Straight line graphs

By the end of this section you will know how to:

* Draw and interpret information from straight line graphs

GCSE LINKS
AF: 22.1 Interpreting and drawing the graphs you meet in everyday life; **BF:** Unit 2 12.1 Interpreting and drawing the graphs you meet in everyday life; **16+:** 8.4 Interpret and draw graphs you meet in everyday life

Key points

* In straight line graphs of real-life situations, you can interpret the relevance of the gradient and the point(s) where the graph crosses the axes.
* You can work out the value of the gradient of a straight line graph using the method shown in Chapter 7.

Guided

1 An approximate conversion rate between imperial pints (pt) and litres is 1 pt = 0.6 litres.

You should know
You need to join the plotted points with a straight line.

a Use the grid to draw a conversion graph to convert between pints and litres.

Pints	0	10	20	30	40	50
Litres	0	6	....	....	....	....

1 pt = 0.6 litre

10 pt = 10 × 0.6 = 6 litres

20 pt = 20 × 0.6 = ... litres

30 pt = 30 × 0.6 = ... litres

40 pt = 40 × .... = ... litres

50 pt = ... = ... litres

Remember this
First set up a table to work out the corresponding numbers of litres equivalent to 10, 20, 30, 40 and 50 pints.

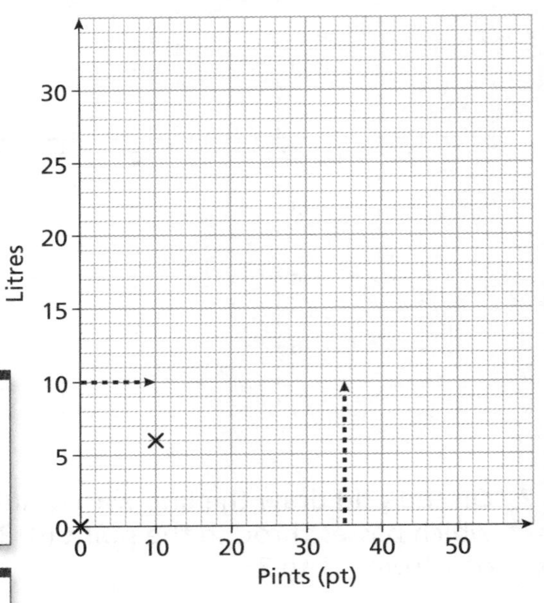

b Use your graph to convert 35 pints to litres.

35 pints = ... litres

Hint
Draw a line up from 35 on the pints axis to meet the line and a line across to the litres axis to read off the number of litres.

c Use your graph to convert 10 litres to pints

10 litres = ... pints

Hint
Draw a line across with an arrow from 10 on the litres axis to meet the line and down with an arrow to the pints axis to read off the number of pints.

Practice

2 The conversion between miles and kilometres (km) is 1 mile = 1.6 km.

a Use the grid to draw a conversion graph to convert between miles and kilometres.

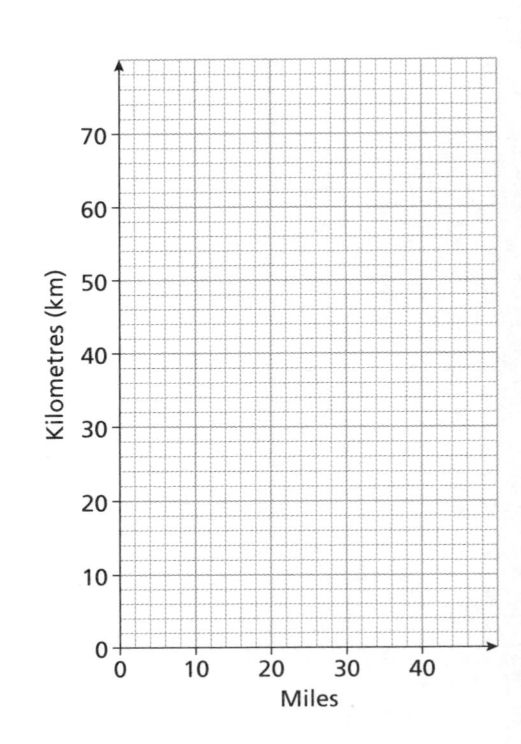

b Use your graph to convert 15 miles to kilometres.

c Use your graph to convert 40 kilometres to miles.

Guided

3 A taxi company charges a standard fee of £4 plus £2 for each mile travelled.

a Use the grid to draw a graph for the cost of a taxi journey.

Journey distance (miles)	0	5	10	15	20	25
Cost in pounds (£)	4	14	....	....	....	....

0 miles: cost = £4

5 miles: cost = 4 + 5 × 2 = £14

10 miles: cost = 4 + 10 × 2 = £........

15 miles: cost = 4 + 15 × .... = £........

20 miles: cost = 4 + ... = £........

25 miles: cost = ... = £........

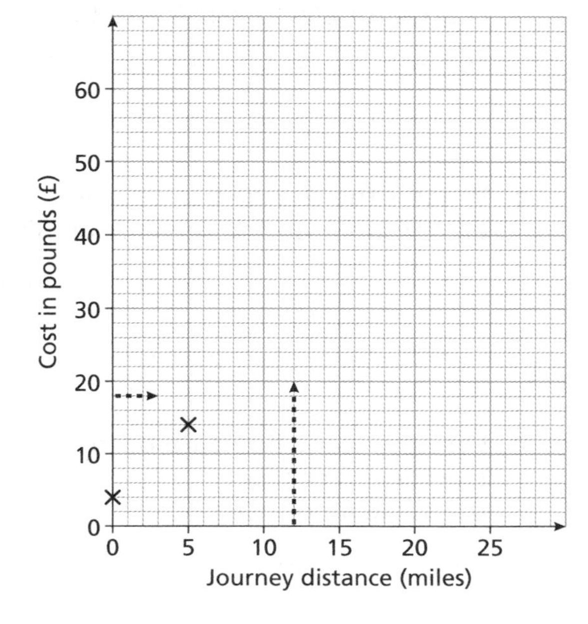

b Use your graph to work out the cost of a taxi journey of 12 miles.

Cost of 12 mile taxi journey = £........

c A taxi journey costs £18. How many miles was the journey?

Journey costing £18 = ... miles

Practice

4 A car hire company charges a basic fee of £30 plus £25 per day.

a Use the grid to draw a graph for the cost of hiring a car for up to 10 days.

Hint
Work out what each small square represents.

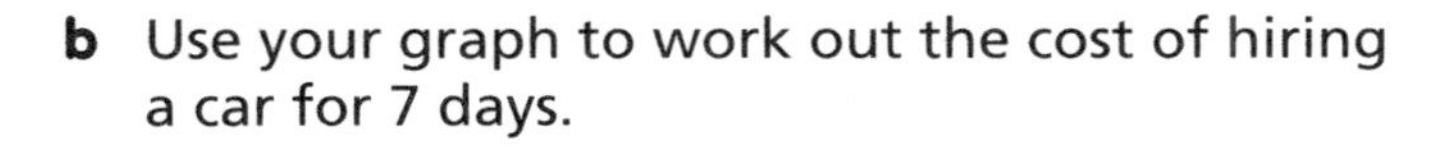

b Use your graph to work out the cost of hiring a car for 7 days.

c The company charged a customer £255 to hire a car.
How many days did the customer hire the car for?

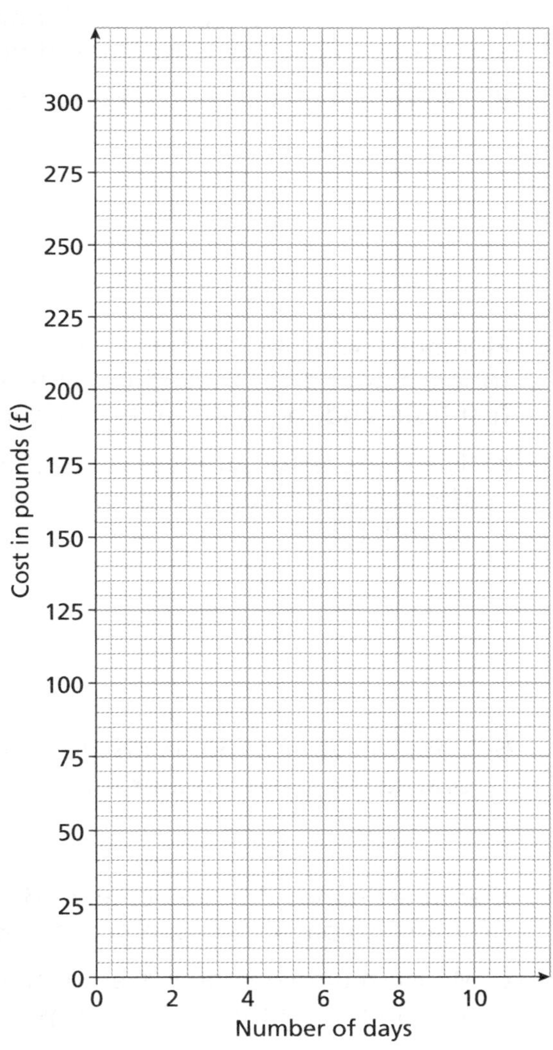

Guided

5 The graph shows the costs charged by two different electricians, A and B.
Both have an initial call-out charge plus a cost per hour.

You should know

The initial call-out charge is the amount charged for 0 hours.
The charge per hour is the gradient of the straight line.

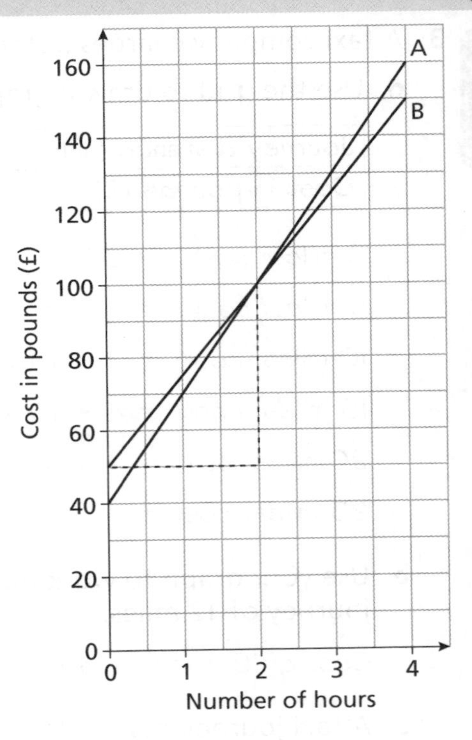

a What is the initial call-out charge for electrician A and for electrician B?

Call-out charge for electrician A is £40

Call-out charge for electrician B is £.......

b What is the charge per hour for electrician A and for electrician B?

Charge per hour for electrician A is $\frac{30}{1}$ = £30/hour

Charge per hour for electrician B is $\frac{\dots\dots}{\dots\dots}$ = £....... /hour

c For how many hours do both electricians charge the same amount?

Number of hours = ...

Hint

The electricians charge the same amount where the straight lines cross.

Kerry estimates the job she needs doing will take 4 hours.
She wants to pay the least amount possible for the job.

d Which electrician should Kerry choose? You must give the reasons for your answer.

From the graph, for 4 hours:

electrician A charges £160 electrician B charges £.......

so Kerry should choose electrician ....

Practice

6 Plumber A charges £50 for the first hour and then £30 for each hour after that. Plumber B charges £30 for the first hour and then £40 for each hour after that.

a The charges for plumber A are shown on the grid. Use the grid to draw a graph of the charges for plumber B.

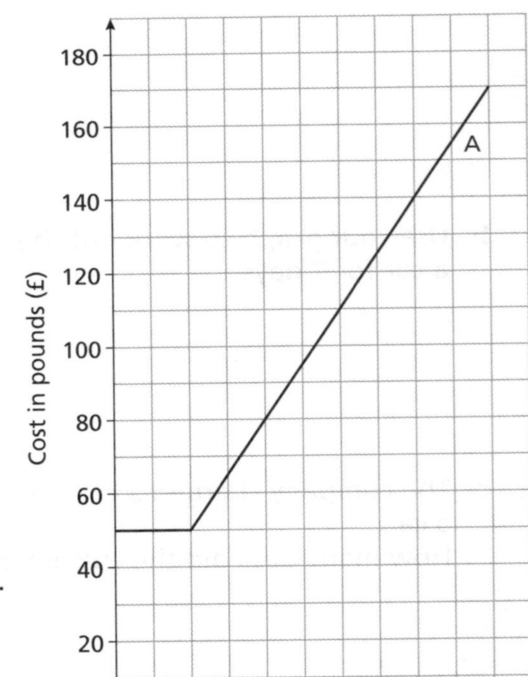

b What number of hours do plumber A and plumber B charge the same amount for?

Kieran estimates the job he needs doing will take 4 hours.
He wants to pay the least amount possible for the job.

c Which plumber should Kieran choose?
You must give the reasons for your answer.

7 The graph shows the cost of attending yoga sessions at three different gyms in any one month.

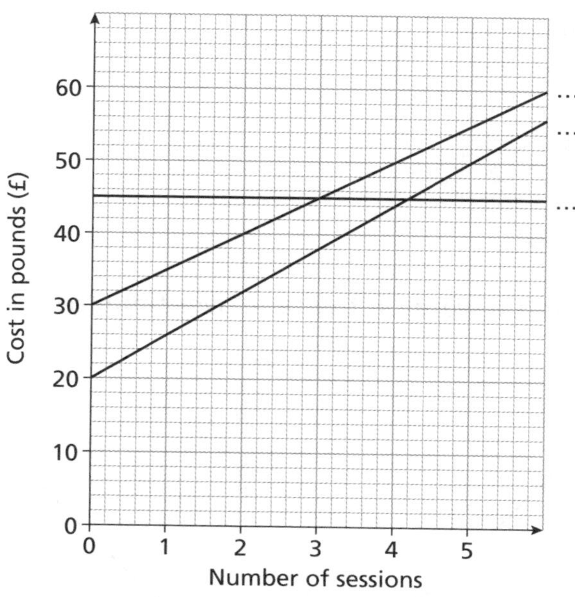

Gym A £45 for any number of sessions

Gym B £30 joining fee plus £5 per session

Gym C £20 joining fee plus £6 per session

a Label each line on the graph with the gym it represents.

b Max wants to attend four yoga sessions per month.
Which gym would be cheapest for him to use? How much would this cost?

Gym ....

£ ....

c Give Helen some advice based on costs about how to choose which gym she should attend for yoga sessions. You must give the reasons for your answer.

....

....

....

Needs more practice ☐ Almost there ☐ I'm proficient! ☐

9.2 Different graph shapes

GCSE LINKS
AH: 15.6 Real-life graphs;
BH: Unit 2 9.6 Real-life graphs

By the end of this section you will know how to:

* Identify graphs of real-life situations

Key points

* Graphs of real-life situations can be used to show how one unit changes in relation to another.
* If the gradient of the graph is positive, as one unit increases the other increases.
* If the gradient of the graph is negative, as one unit increases the other decreases.

Remember this

When a container is being filled at a constant rate, the wider the container, the slower the height will increase with time. The narrower the container, the quicker the height will increase with time, i.e. the gradient of the graph will be steeper.

Guided

1 Water is emptied out of a large fish tank in the shape of a cuboid.

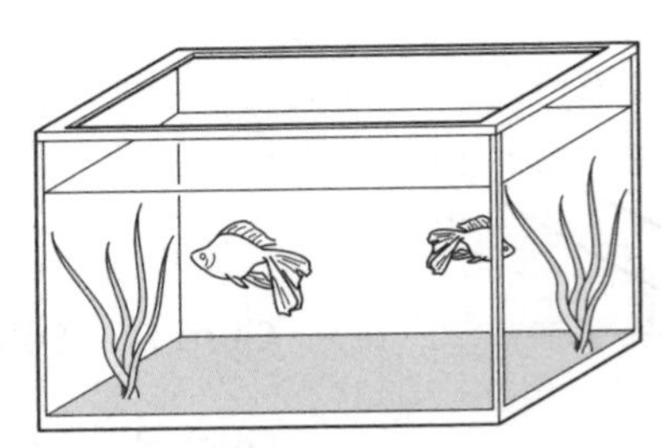

Which sketch graph could represents the height (h) of the water in the tank over time (t)?

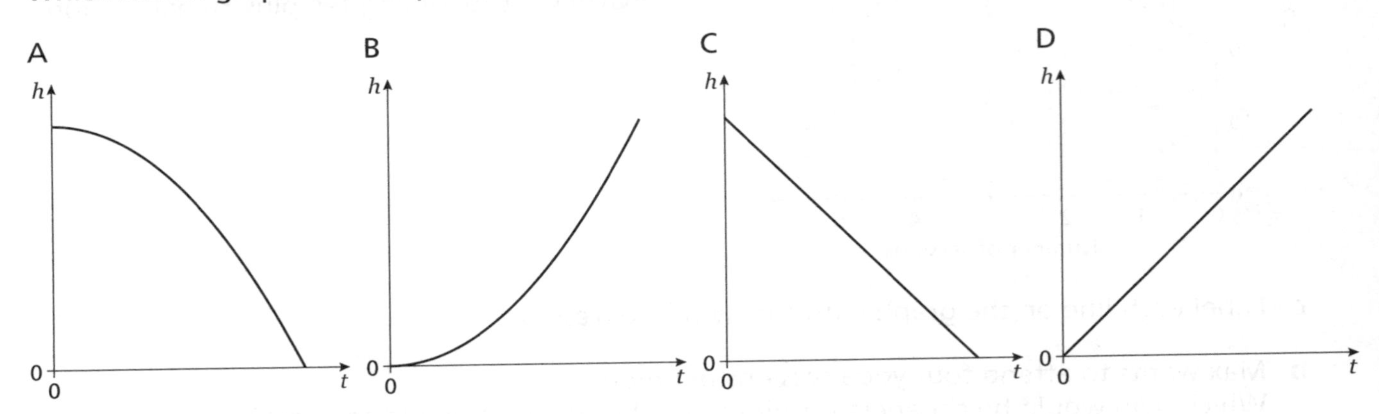

The fish tank starts off full of water so graphs B and D must be incorrect. (In these graphs the height starts off as 0.)
The fish tank is the shape of a cuboid so the water will empty at a steady pace. Answer: graph C is correct

Remember this

In a cuboid, the height of the remaining water will decrease by the same amount per second so the graph of height against time will be a straight line with constant gradient.

Practice

2 Liquid is poured into each of these containers at a steady rate.

A

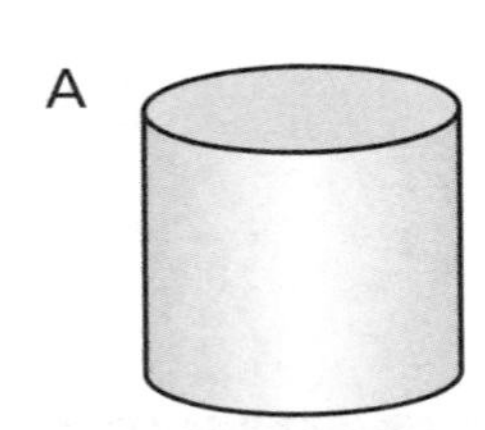

B

C

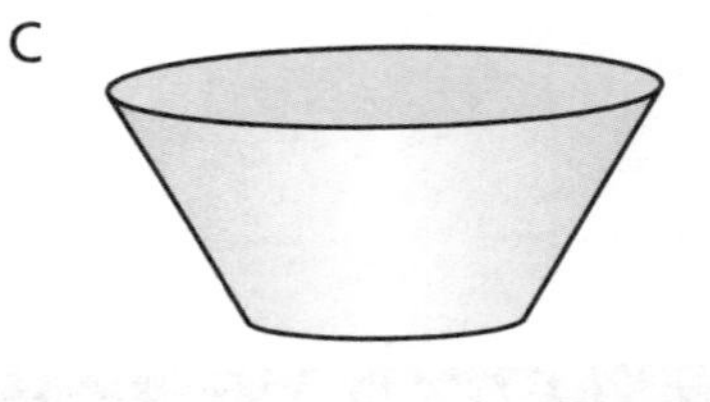

Match each container with the correct graph showing the height of the liquid in the container over time. Write each container letter under the correct graph.

1 Height (cm), Time (s)

2 Height (cm), Time (s)

3 Height (cm), Time (s)

....

Remember this

If the container gets narrower at the top, then the height of the water increases more rapidly per second, so the gradient of the graph will get steeper. If the container gets wider at the top, the height of the water increases more slowly per second, so the gradient will become less steep.

3 Water is poured into each of these different-shaped vases at a steady rate.

A

B

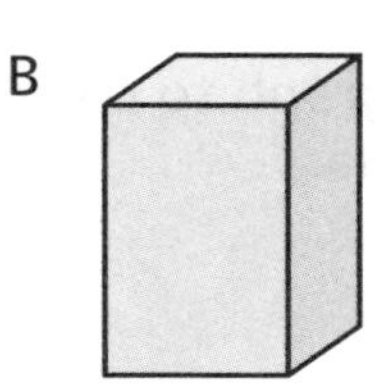

C

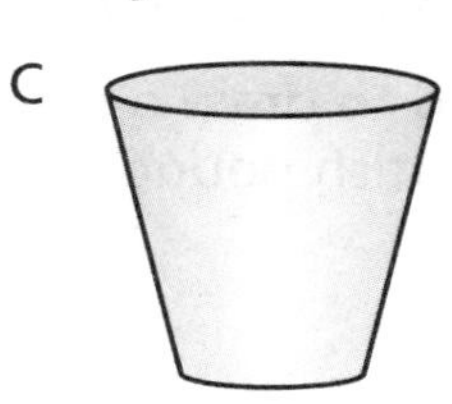

For each vase, sketch a graph to show the height (h) of water in the vase over time (t).

Don't forget!

* In straight line graphs of real-life situations, you can interpret the relevance of the .... and the point(s) where the graph crosses the ....
* If the gradient of a graph is positive, as one unit increases the other ....
* If the gradient of a graph is negative, as one unit increases the other ....
* Match each graph to two of the statements.

A

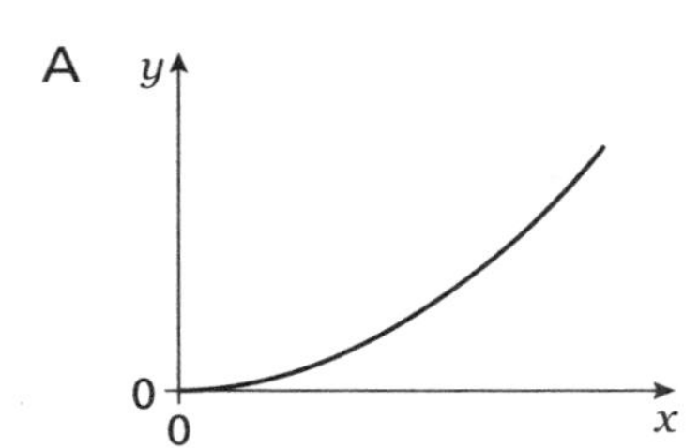

B

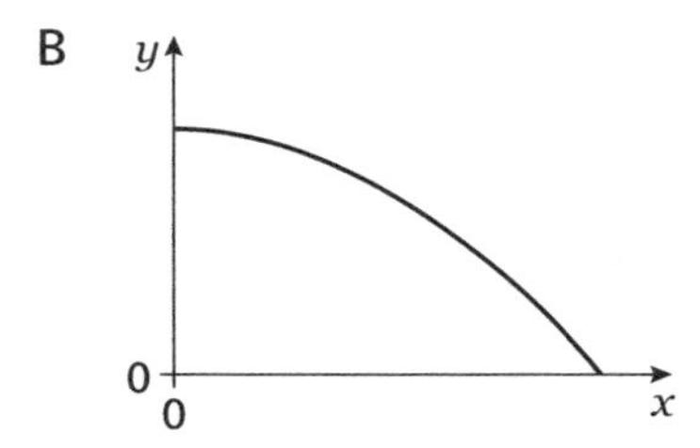

negative gradient	positive gradient	as one unit increases, the other increases	as one unit increases, the other decreases

Exam-style questions

1 The graph shows the cost of posting a small packet.

a How much does it cost to post a packet weighing 50 g?

£ ....

b A customer is charged £3.60 to post a packet.
What is the smallest possible weight of the packet?

Cost in pounds (£)

Weight in grams (g)

.... grams

2 An approximate conversion from British pounds (£) to US dollars ($) is £1 = $1.60.

a Use the grid to draw a conversion graph to convert between British pounds and US dollars.

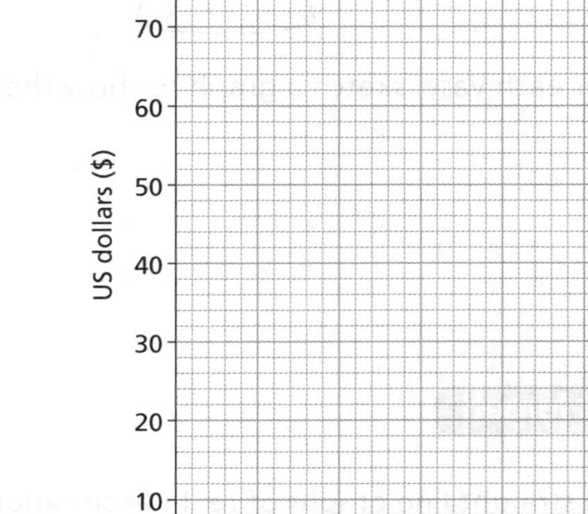

b Use your graph to convert £35 to US dollars.

$ ....

c Use your graph to convert $64 to British pounds.

£ ....

3 A hot cup of coffee is left in a room to cool.
Which of the following graphs represents the temperature in °C of the cup of coffee over time?

A

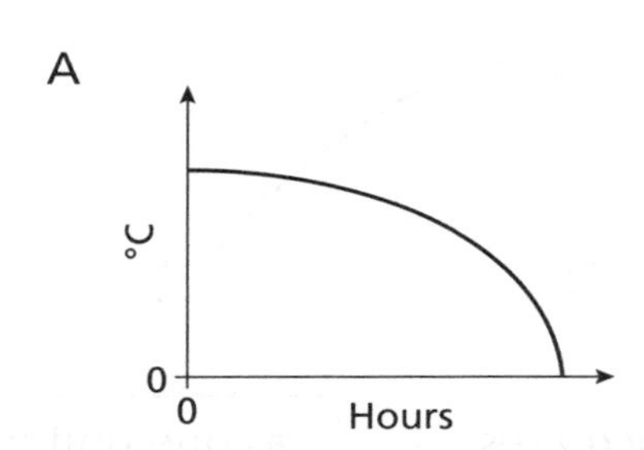

B

°C
0
0
Hours

C

°C
0
0
Hours

D

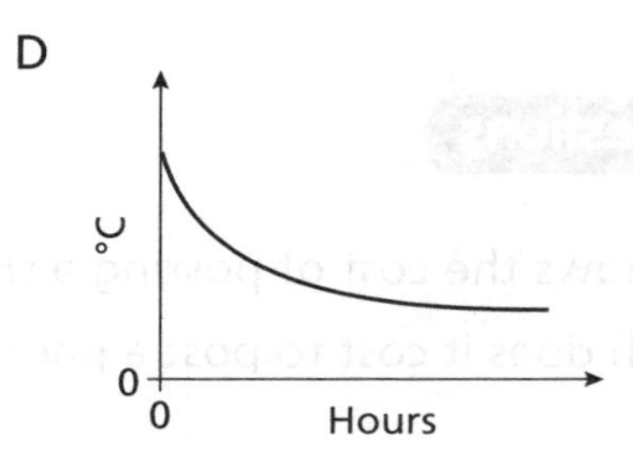

4 The graph shows the height of a lighted candle over time.

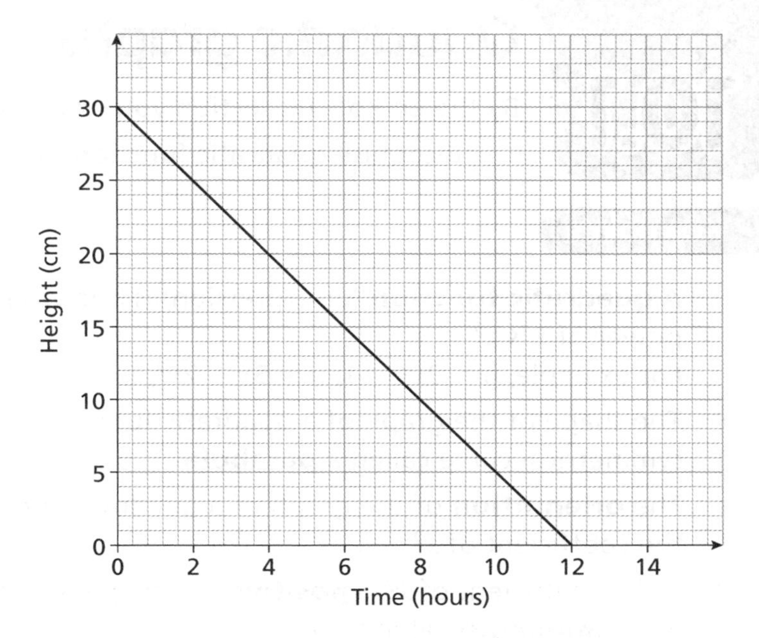

a What is the height of the candle at the start?

.... cm

b How long does it take the candle to burn down to half its height?

.... hours

c What is the height of the candle after 3 hours?

.... cm

5 The graph shows the charges of three different car hire companies.
The charges of the three car hire companies are

Car hire A	£40 basic fee plus £30 per day
Car hire B	£60 plus £25 per day
Car hire C	£120 total cost for days 1, 2 and 3 then £35 per day

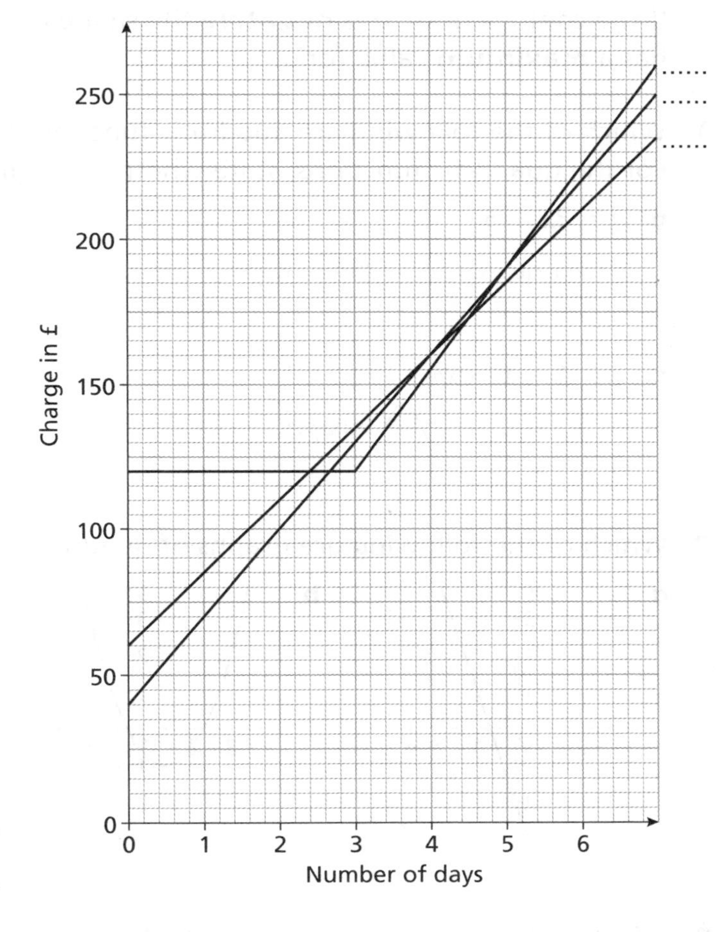

a Label each line on the graph with the car hire company it represents.

b Bilal wants to hire a car for 2 days.
Which car hire company would be cheapest for him to use?
How much would this cost?

Company ....

£ ....

c Ali wants to hire a car for 4 days.
Which car hire company would be cheapest for her to use?
You must give the reasons for your answer.

....

....

Needs more practice ☐ Almost there ☐ I'm proficient! ☐

10.1 Quadratic graphs

By the end of this section you will know how to:

* Recognise quadratic functions and their graphs

GCSE LINKS

AH: 21.1 Graphs of quadratic functions; **BH:** Unit 3 6.1 Graphs of quadratic functions; **16+:** 8.6 Draw and use graphs of quadratic functions to solve equations

Key points

* A **quadratic function** (or expression) is one where the highest power of x is x^2. For example, $y = 3x^2 - 1$, $y = 5x - 2x^2$, $y = 6x^2$, $y = 2x^2 + 3x + 1$.
* The shape of the graph of a quadratic function is a smooth, symmetrical curve called a **parabola**.
* The **orientation** of the graph of a quadratic function depends on the coefficient of x^2.
* If the coefficient of x^2 is **positive** the shape is ∪ and the graph has a **minimum** value for y.
* If the coefficient of x^2 is **negative** the shape is ∩ and the graph has a **maximum** value for y.

You should know

The **coefficient** of x^2 is the number in front of the x^2. For example, in $y = 5x^2 - 3x + 4$ the coefficient of x^2 is 5.

Remember this

Orientation means direction, i.e. in the case of a quadratic graph whether the curve points up or down.

Guided

1 Which of the following are quadratic functions? For the quadratic functions, state whether the graph is a ∪ shape or a ∩ shape.

a $y = 3x - 5$

Not a quadratic as highest power is x.

b $y = 2x^2 + 3$

Quadratic and ∪ shape as coefficient of x^2 is ....

c $y = 4 - x - x^2$

Quadratic and ∩ shape as coefficient of x^2 is ....

d $y = x^3 - 4x^2$

Not a quadratic as highest power is x^3.

Practice

2 Which of the following are graphs of quadratic functions? ....

a

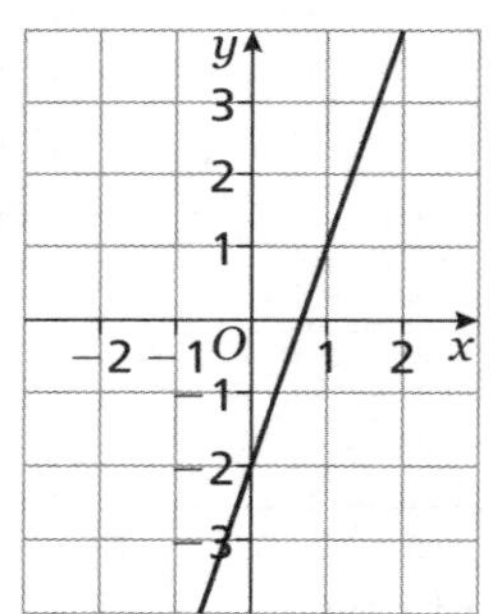

b

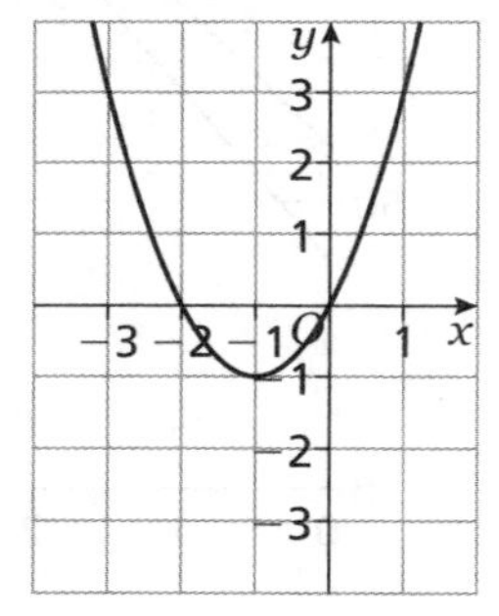

c

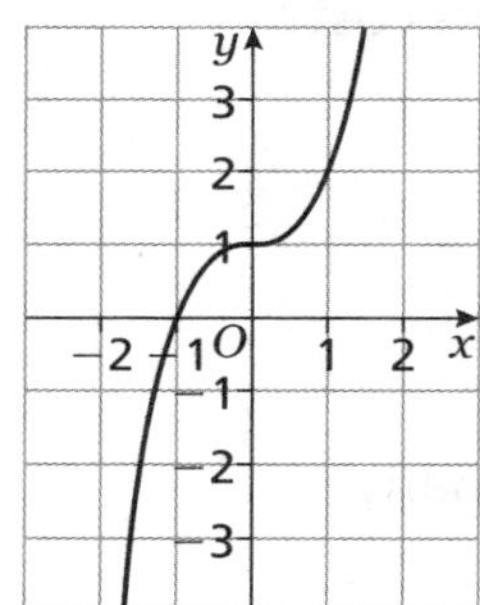

d

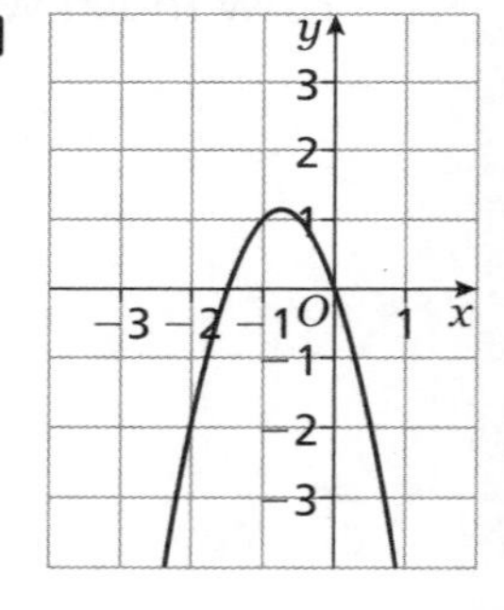

3 Which of the following are quadratic functions? For the quadratic functions, sketch the shape of the graph.

a $y = 5 - x^2$

b $y = 7x + 4$

c $y = 4x^2 - 7x$

d $y = 3x^3 - 6x^2$

4 Which of the following are graphs of quadratic functions? ....

a

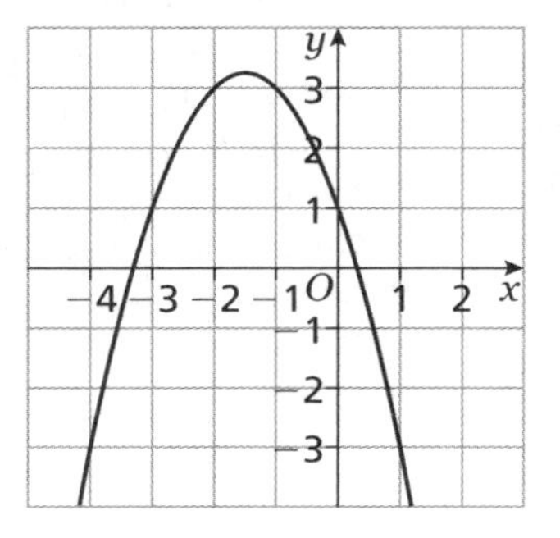

b

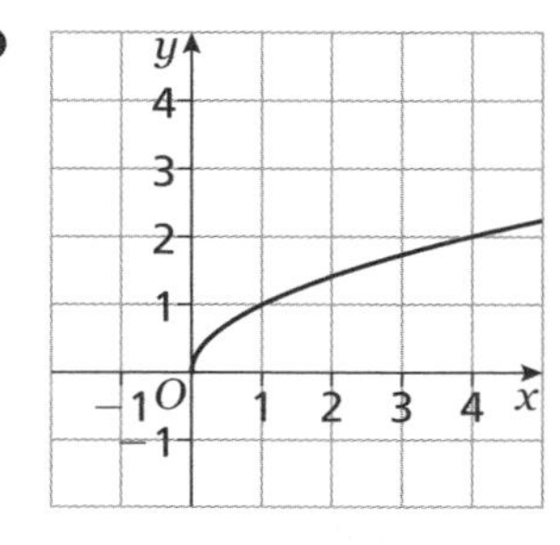

c

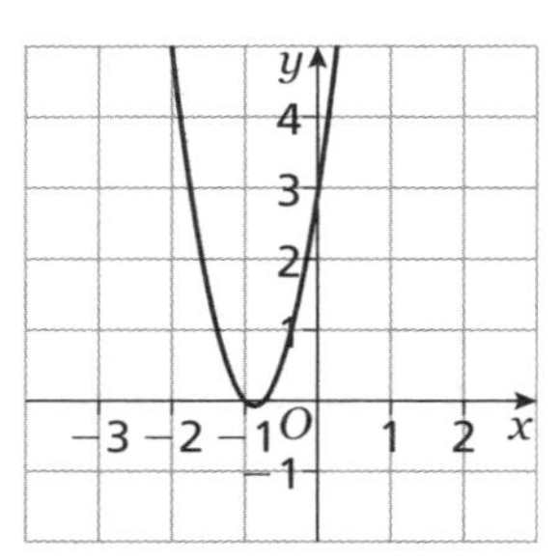

d

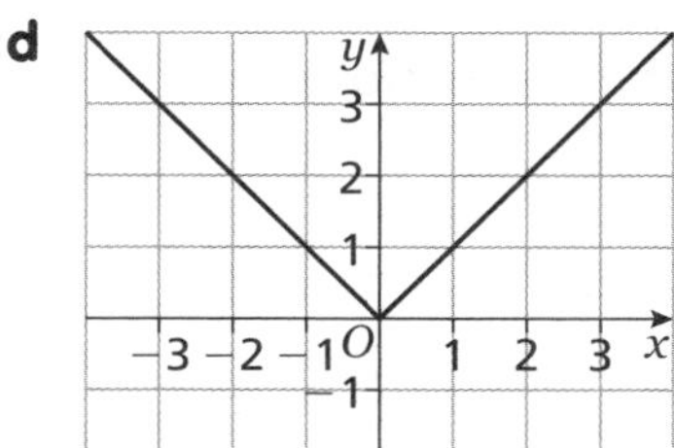

Needs more practice ☐ Almost there ☐ I'm proficient! ☐

10.2 Graphs of the form $y = ax^2$

By the end of this section you will know how to:

* Sketch graphs of the form $y = ax^2$

GCSE LINKS

AH: 21.1 Graphs of quadratic functions; **BH:** Unit 3 6.1 Graphs of quadratic functions; **16+:** 8.6 Draw and use graphs of quadratic functions to solve equations

Key points

* For quadratic graphs of the form $y = ax^2$:
  - the y-axis is a line of symmetry
  - the graph passes through the point with coordinates (0, 0)
  - if a is **positive** the graph has a **minimum** point at (0, 0)
  - if a is **negative** the graph has a **maximum** point at (0, 0)
* When **sketching** a graph, it is not necessary to draw it on graph paper. Label the axes and make sure the shape and the point where the graph crosses the y-axis are clearly shown.

Guided

1 **a** Sketch the graph of $y = x^2$.

The graph is ∨ shaped because the coefficient of x^2 is ....

The graph passes through the point (0, 0).

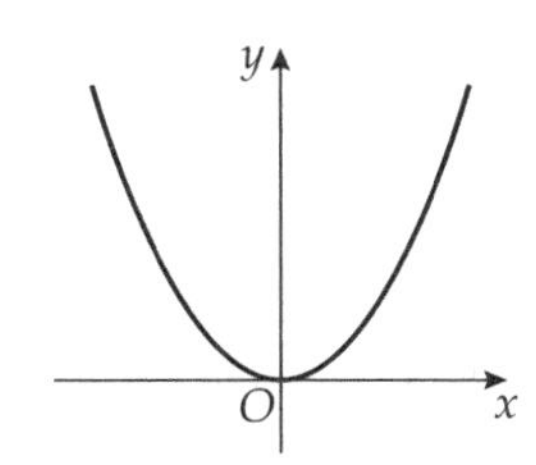

Remember this

The graph of $y = x^2$ is above the x-axis. As the value of x gets bigger the curve gets steeper and the value of y increases more quickly. The same happens as the value of x gets smaller.

b What is the minimum value of y on the graph?

The minimum value of y is 0.

Practice

2 a Sketch the graph of $y = -x^2$.

Hint
To find the y values for the graph of $y = -x^2$ multiply all the values of y in the graph of $y = x^2$ by -1. The graph of $y = -x^2$ is a reflection of the graph of $y = x^2$ in the x-axis.

Remember this
The graph of $y = -x^2$ is below the x-axis. As the value of x gets bigger the curve gets steeper and the value of y decreases more quickly. The same happens as the value of x gets smaller.

b What is the maximum value of y on the graph? ....

Guided

3 Sketch the graphs of $y = x^2$ and $y = 3x^2$ on the same axes.

Both graphs are .... shaped as the coefficient of x^2 in both equations is ....

Both graphs pass through the point (......, ...).

To find the y values for the graph of $y = 3x^2$, multiply each of the values of y from the graph of $y = x^2$ by ....

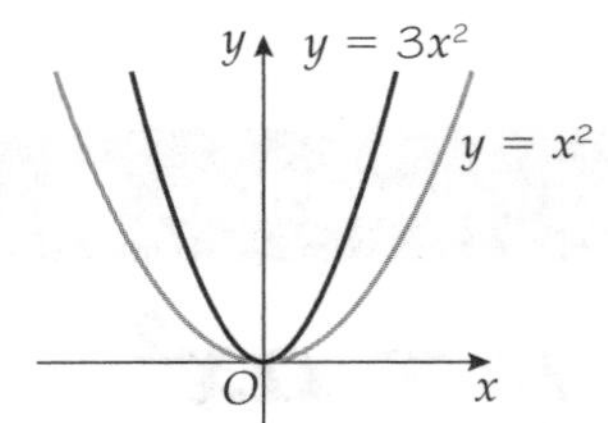

Remember this
The graph of $y = 3x^2$ is steeper on both sides than the graph of $y = x^2$ so the curve is narrower.

Practice

4 Sketch the graphs of $y = -x^2$ and $y = -4x^2$ on the same axes.

Needs more practice ☐ Almost there ☐ I'm proficient! ☐

10.3 Graphs of the form $y = ax^2 + b$

By the end of this section you will know how to:

* Sketch graphs of the form $y = ax^2 + b$

GCSE LINKS
AH: 21.1 Graphs of quadratic functions; **BH:** Unit 3 6.1 Graphs of quadratic functions; **16+:** 8.6 Draw and use graphs of quadratic functions to solve equations

Key points

* For quadratic graphs of the form $y = ax^2 + b$:
  - b is the number where the graph crosses the y-axis, i.e. the graph passes through (0, b)
  - the y-axis is a **line of symmetry**
  - if a is **positive** the graph has a **minimum** point at (0, b)
  - if a is **negative** the graph has a **maximum** point at (0, b)
  - the graph of $y = ax^2 + b$ is a **translation** of the graph of $y = ax^2$ by b units in the y-axis.
* In examples where the graph crosses the x-axis you may be asked to identify these points.

1 a Sketch the graph of $y = x^2 - 9$.

The graph is ∪ shaped, as the coefficient of x^2 is +.

The graph crosses the y-axis when $x = 0$, so $y = 0 - 9 =$ ....

The graph passes through the point (0, ...).

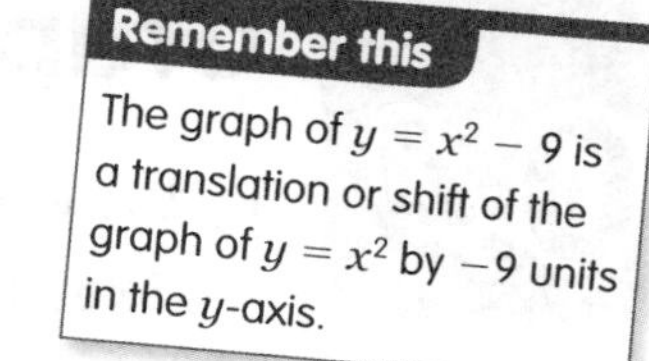
Remember this
The graph of $y = x^2 - 9$ is a translation or shift of the graph of $y = x^2$ by -9 units in the y-axis.

b What is the minimum value of y on the graph?

The minimum value of y is ....

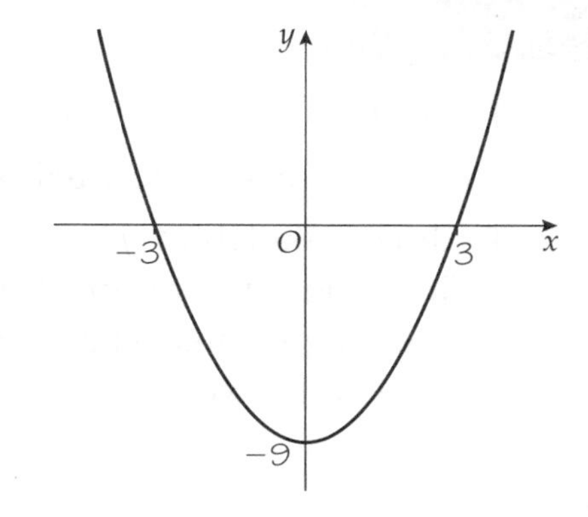

c Where does the graph cross the x-axis?

$0 = x^2 - 9$

$x^2 - 9 = 0$

$x^2 = 9$

$x = \pm\sqrt{9}$ so $x = 3$ and $x = -3$

Hint
The graph crosses the x-axis when $y = 0$.

2 a Sketch the graph of $y = -x^2 + 16$.

b What is the maximum value of y on the graph? ....

c Where does the graph cross the x-axis? ....

3 a Sketch the graph of $y = -2x^2 - 3$.

The graph of $y = -2x^2$ is steeper than the graph of $y = -x^2$.

To find the y values for the graph of $y = -2x^2$,

multiply the y values from the graph of $y = -x^2$ by ....

The graph of $y = -2x^2 - 3$ is a translation of the graph of

$y = -2x^2$ by .... units in the y-axis.

The graph will pass through the point (0, ...).

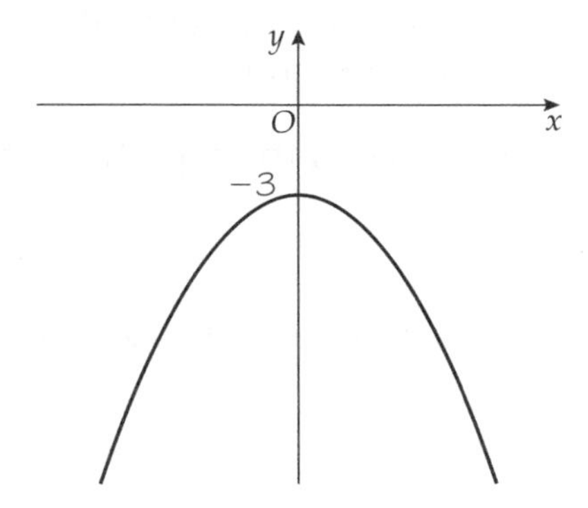

b What is the maximum value of y on the graph?

The maximum value of y is ....

4 Sketch the graph of $y = 3x^2 - 3$. You must label relevant information on the sketch, including the points where the graph crosses the x-axis.

Remember this
The graph of $y = ax^2 + b$ is a translation of the graph of $y = ax^2$ by b units in the y-axis.

10.4 Graphs of the form $y = (x + b)^2$

By the end of this section you will know how to:

* Sketch graphs of the form $y = (x + b)^2$

GCSE LINKS
AH: 21.1 Graphs of quadratic functions; **BH:** Unit 3 6.1 Graphs of quadratic functions; **16+:** 8.6 Draw and use graphs of quadratic functions to solve equations

Key points

* The graph of $y = (x + b)^2$ is a ∨ shaped curve.
* The graph meets the x-axis at one point.
* To draw a sketch of $y = (x + b)^2$:
  - find where the graph crosses the y-axis by putting $x = 0$
  - find where the graph meets the x-axis by putting $y = 0$
  - label all the key points on your sketch.

Guided

1 For the graph of $y = (x + 2)^2$:

a Find the point where the graph crosses the y-axis.

$y = (0 + 2)^2$

$y = 2^2$

$y = 4$ so curve passes though (0, 4)

Hint
The graph crosses the y-axis when $x = 0$.

b Find the point where the graph meets the x-axis.

$0 = (x + 2)^2$

$(x + 2)^2 = 0$

$x + 2 = 0$

$x = -2$ so curve meets the x-axis at $(-2, 0)$

Hint
The graph meets the x-axis when $y = 0$.

c Sketch the graph of $y = (x + 2)^2$ labelling all the key points on the graph.

d What is the equation of the line of symmetry of the graph?

$x =$ ....

e What are the coordinates of the minimum point?

$(-2,$ ....)

Remember this
The graph of $y = (x + 2)^2$ is a translation of the graph of $y = x^2$ by 2 units to the left.

2 For the graph of $y = (x - 3)^2$:

a Find the point where the graph crosses the y-axis.

$y = (0 - 3)^2$

$y = (-3)^2$

$y =$ .... so curve passes though (0, ...)

Hint
The graph crosses the y-axis when $x = 0$.

b Find the point where the graph meets the x-axis.

$0 = (x - 3)^2$

$(x - 3)^2 = 0$

$x - 3 = 0$

$x =$ .... so curve meets the x-axis at (......., 0)

Hint
The graph meets the x-axis when $y = 0$.

c Sketch the graph of $y = (x - 3)^2$ labelling all the key points on the graph.

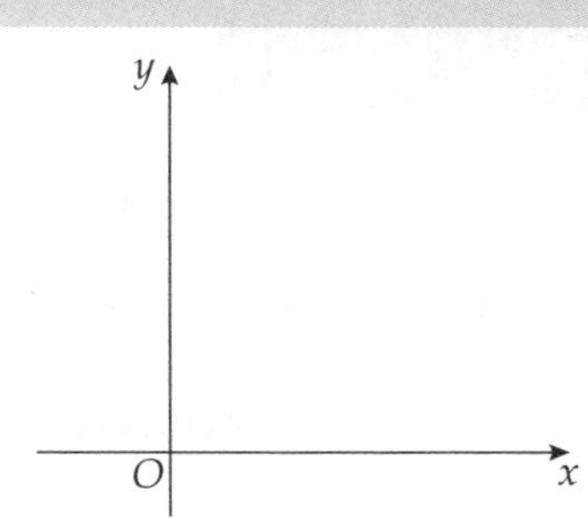

d What is the equation of the line of symmetry of the graph? $x =$ ....

e What are the coordinates of the minimum point? (......., ...)

Practice

3 For the graph of $y = (x + 4)^2$:

a Find the point where the graph crosses the y-axis.

b Find the point where the graph meets the x-axis.

c Sketch the graph of $y = (x + 4)^2$ labelling all the key points on the graph.

d What is the equation of the line of symmetry of the graph?

e What are the coordinates of the minimum point?

4 For the graph of $y = (x - 6)^2$:

a Find the point where the graph crosses the y-axis.

b Find the point where the graph meets the x-axis.

c Sketch the graph of $y = (x - 6)^2$ labelling all the key points on the graph.

d What is the equation of the line of symmetry of the graph?

e What are the coordinates of the minimum point?

Don't forget!

- A quadratic function (or expression) is one where the highest power of x is ....
- The shape of the graph of a quadratic function is symmetrical and is called a ....
- The orientation of the graph of a quadratic function depends on the coefficient of ....
- If the coefficient is .... the curve is ∪ shaped.
- If the coefficient is .... the curve is ∩ shaped.
- A quadratic graph has a line of ....
- A graph meets the y-axis when ....
- A graph meets the x-axis when ....
- In the graph of $y = 3x^2$:
  - the graph crosses the y-axis at the point with coordinates (......, ...)
  - the minimum value of y is ......
  - the y values for the graph of $y = 3x^2$ can be found from the y values of the graph of $y = x^2$ by .... by 3.
- In the graph of $y = -x^2 + 5$:
  - the graph crosses the y-axis at the point with coordinates (......, ...)
  - the maximum value of y is ......
  - the graph of $y = -x^2 + 5$ is a translation of the graph of $y = -x^2$ by ...... units in the ...... axis.
- The graph of $y = ax^2 + b$ is a translation of the graph of $y = ax^2$ by ...... units in the y-axis.
- In the graph of $y = (x + 3)^2$:
  - the graph crosses the y-axis at the point with coordinates (......, ...)
  - the graph meets the x-axis at the point with coordinates (......, ...)
  - the equation of the line of symmetry is ....
  - the graph of $y = (x + 3)^2$ is a translation of the graph of $y = x^2$ by ...... units to the ....

Exam-style questions

1 **a** Sketch the graph $y = x^2 + 7$
You must label relevant information on the graph.

b What is the minimum value of y on the graph?

....

2 Sketch the graphs of $y = x^2$ and $y = 5x^2$ on the same axes.
You must label relevant information on the graphs.

3 **a** Sketch the graph $y = -3x^2 + 12$
You must label relevant information on the graph,
including the points where the graph crosses the x-axis.

b What is the maximum value of y on the graph?

....

4 **a** Sketch the graphs of $y = -x^2$ and $y = -2x^2 - 4$ on the same axes.
You must label relevant information on the graphs.

b What is the maximum value of y on the graph of $y = -x^2$?

....

c What is the maximum value of y on the graph of $y = -2x^2 - 4$?

....

5 For the graph of $y = (x - 5)^2$

a Find the point where the graph crosses the y-axis.

....

b Find the point where the graph meets the x-axis.

....

c Sketch the graph of $y = (x - 5)^2$ labelling all the key points on the graph.

d What is the equation of the line of symmetry of the graph?

....

Needs more practice ☐ Almost there ☐ I'm proficient! ☐

11.1 Plot graphs of quadratic functions of the form $y = ax^2 + b$

By the end of this section you will know how to:

* Use a table of values to plot the graph of a quadratic function of the form $y = ax^2 + b$

GCSE LINKS
AH: 21.1 Graphs of quadratic functions; **BH:** Unit 3 6.1 Graphs of quadratic functions; **16+:** 8.6 Draw and use graphs of quadratic functions

Key points

* To plot a quadratic graph you need to set up a table of values for x.
* You can find the corresponding values of y by substituting values for x into the quadratic equation.
* After plotting the points, join them with a smooth curve.
* For graphs of functions of the form $y = ax^2 + b$, the y-axis is a line of symmetry.

Remember this
x^2 is always greater than or equal to 0, i.e. a number squared cannot be negative, e.g. $2^2 = 4$ and $(-3)^2 = 9$.

Guided

1 a Complete the table of values for $y = x^2$.

x	−3	−2	−1	0	1	2	3
$y = x^2$	9	4	....	....	....	....	....

$x = -3$: $y = (-3)^2 = 9$ $x = -2$: $y = (-2)^2 = 4$

$x = -1$: $y = (-1)^2 =$ .... $x = 0$: $y = ($.......$)^2 =$ ....

$x = 1$: $y = ($.......$)^2 =$ .... $x = 2$: $y =$ .... $=$ ....

$x = 3$: $y =$ .... $=$ ....

b On the grid draw the graph of $y = x^2$ for values of x from −3 to +3.

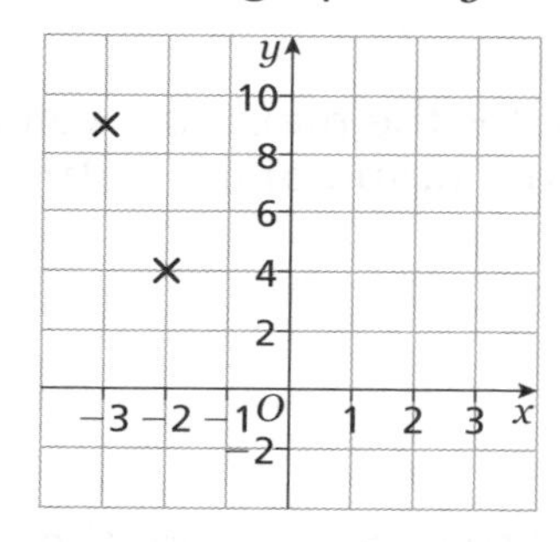

Remember this
Check that your graph is the correct shape. For $y = x^2$ the coefficient of x^2 is positive (+1) so the curve should be ⋃ shaped. Try to make sure the bottom of the curve is curved and not pointed.

2 On the grid draw the graphs of $y = -x^2$ and $y = -2x^2$ for values of x between −3 and +3.

You should know
$-x^2 \neq (-x)^2$

x	−3	−2	−1	0	1	2	3
$y = -x^2$	−9	−4	....	....	....	....	....

x	−3	−2	−1	0	1	2	3
$y = -2x^2$	−18	−8	....	....	....	....	....

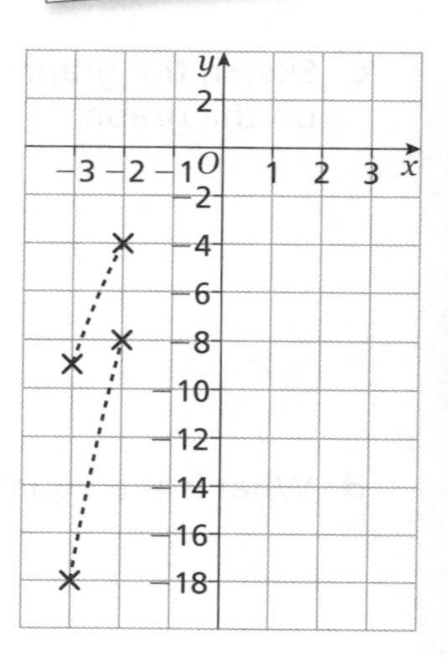

Hint
For the graph of $y = -x^2$ all the values of y are '−' the values you worked out for $y = x^2$.

Hint
For the graph of $y = -2x^2$ multiply all the values you worked out for $y = -x^2$ by 2.

You should know
In the graph of $y = ax^2$, for larger values of a, the graph gets steeper more quickly. If a is negative the graph is ⋂ shaped.

3 On the grid draw the graphs of $y = x^2 + 2$ and $y = x^2 - 3$ for values of x between -3 and $+3$.

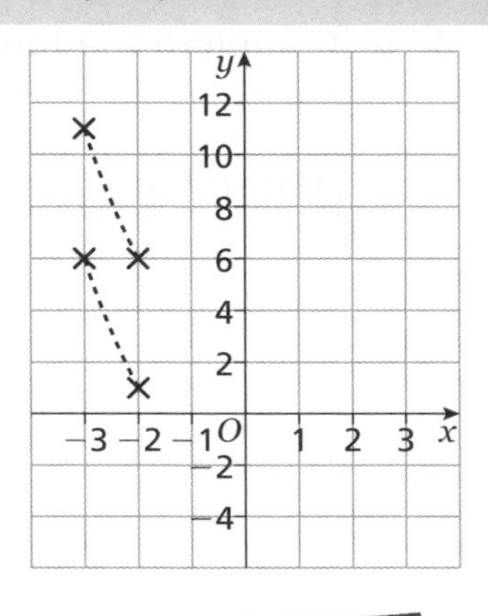

x	−3	−2	−1	0	1	2	3
$y = x^2 + 2$	11	6	....	....	....	....	....

$x = -3$: $y = (-3)^2 + 2 = 11$ $x = -2$: $y = (-2)^2 + 2 = 6$

$x = -1$: $y = (-1)^2 + 2 =$ .... $x = 0$: $y = ($.......$)^2 + 2 =$ ....

$x = 1$: $y = ($.......$)^2 + 2 =$ .... $x = 2$: $y =$ .... $=$ ....

$x = 3$: $y =$ .... $=$ ....

x	−3	−2	−1	0	1	2	3
$y = x^2 - 3$	6	1	....	....	....	....	....

$x = -3$: $y = (-3)^2 - 3 = 6$ $x = -2$: $y = (-2)^2 - 3 = 1$

$x = -1$: $y = (-1)^2 - 3 =$ .... $x = 0$: $y = ($.......$)^2 - 3 =$ ....

$x = 1$: $y = ($.......$)^2 - 3 =$ .... $x = 2$: $y =$ .... $=$ ....

$x = 3$: $y =$ .... $=$ ....

You should know

For graphs of the form $y = ax^2 + b$, b is the number where the graph crosses the y-axis (it is the y-intercept), i.e. the graph passes through the point with coordinates $(0, b)$.

Practice

4 a Complete the table of values for $y = 3x^2$.

x	−3	−2	−1	0	1	2	3
$y = 3x^2$							

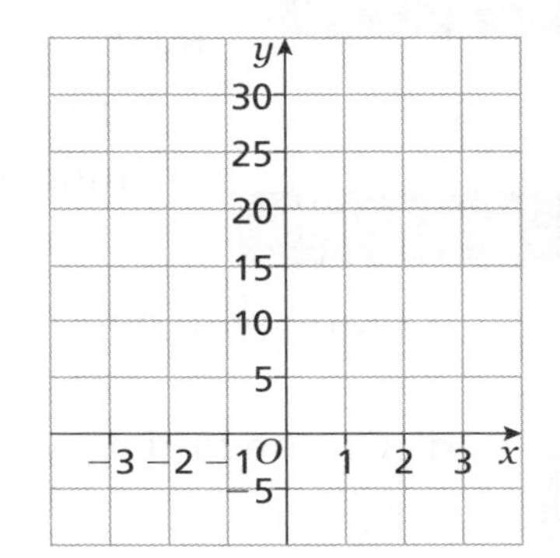

b On the grid draw the graph of $y = 3x^2$ for values of x from -3 to $+3$.

5 On the grid draw the graphs of $y = -x^2 - 1$ and $y = -x^2 + 3$ for values of x between -3 and $+3$.

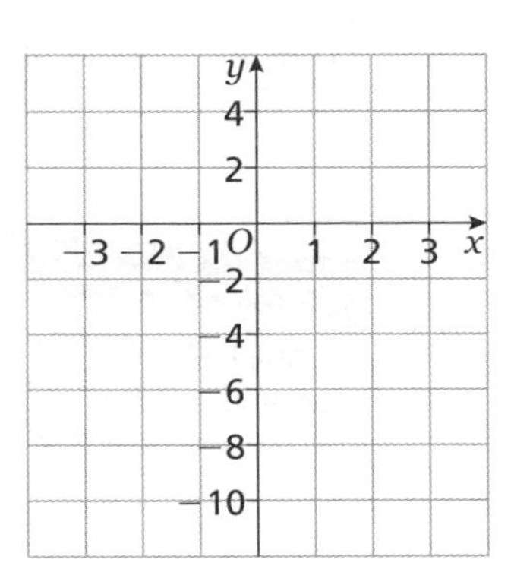

6 **a** Complete the table of values for $y = 4x^2 - 5$.

x	-3	-2	-1	0	1	2	3
$y = 4x^2 - 5$							

b On the grid draw the graph of $y = 4x^2 - 5$ for values of x from -3 to $+3$.

Needs more practice ☐ Almost there ☐ I'm proficient! ☐

11.2 Plot graphs of quadratic functions of the form $y = ax^2 + bx + c$

GCSE LINKS
AH: 21.1 Graphs of quadratic functions; **BH:** Unit 3 6.1 Graphs of quadratic functions; **16+:** 8.6 Draw and use graphs of quadratic functions

By the end of this section you will know how to:

* Use a table of values to plot the graph of a quadratic function of the form $y = ax^2 + bx + c$

Key points

* A graph with equation $y = ax^2 + bx + c$ passes through the point $(0, c)$, i.e. when $x = 0$, $y = c$.

Guided

1 **a** On the grid draw the graph of $y = x^2 - 3x + 4$ for values of x between -2 and $+5$.

x	-2	-1	0	1	2	3	4	5
x^2	4	1	....	1	4	....	16	....
$-3x$	$+6$	....	....	-3	-6	....	-12	....
$+4$	$+4$	$+4$	$+4$	$+4$	$+4$	$+4$	$+4$	$+4$
$y = x^2 - 3x + 4$	14	....	....	2	2	....	8	....

Remember this
You can use a longer version of a table to work out the value of y for each value of x. In $y = x^2 - 3x + 4$, for each column add 4 to the values for x^2 and $-3x$.

You should know
A graph crosses the x-axis when $y = 0$. A graph crosses the y-axis when $x = 0$.

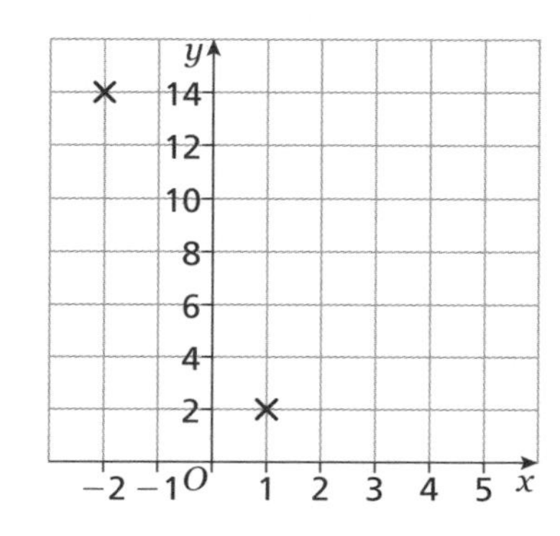

b Draw in and write down the equation of the line of symmetry of the graph.

The line of symmetry passes through the minimum point on the graph.

When $x = 1$, $y = 2$ When $x = 2$, $y = 2$

So minimum point occurs when x is 1.5. Equation is $x = 1.5$

Practice

2 a On the grid draw the graph of $y = x^2 + 2x - 3$ for values of x between -4 and $+2$.

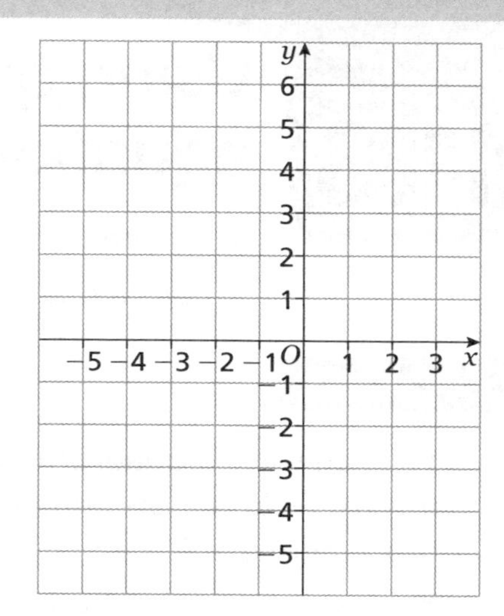

b Write down the equation of the line of symmetry of the graph.

3 a Complete the table of values for $y = x^2 - x + 2$.

x	−2	−1	0	1	2	3	4
y	8			2	4		

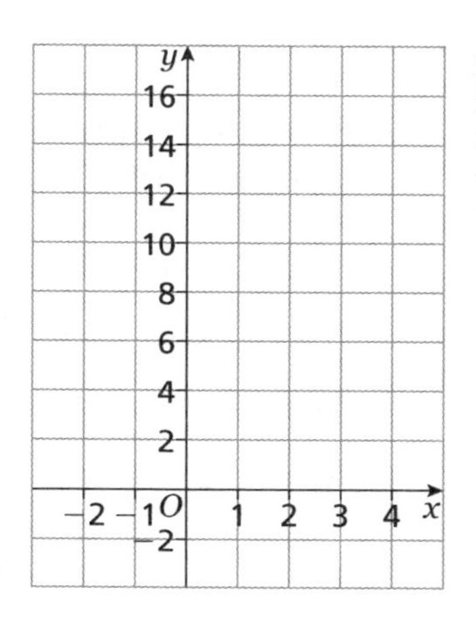

b On the grid draw the graph of $y = x^2 - x + 2$ for values of x between -2 and $+4$.

c Write down the equation of the line of symmetry of the graph.

GCSE

4 a Complete the table of values for $y = x(x - 4)$ for values of x from 0 to 5.

x	0	1	2	3	4	5
y	0	−3			0	

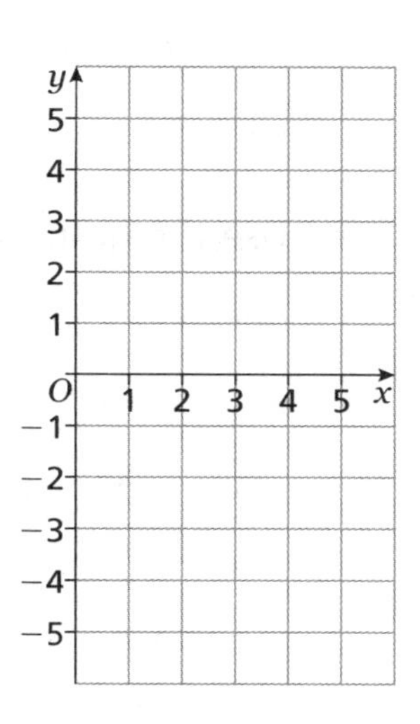

b On the grid draw the graph of $y = x(x - 4)$.

11.3 Using quadratic graphs to solve equations

GCSE LINKS
AH: 21.1 Graphs of quadratic functions; **BH:** Unit 3 6.1 Graphs of quadratic functions

By the end of this section you will know how to:

* Use a quadratic graph to solve an equation

Key points

* You can solve a quadratic equation of the form $ax^2 + bx + c = 0$, by drawing the graph of $y = ax^2 + bx + c$ and finding where the graph crosses the x-axis, i.e. finding the value(s) of x where $y = 0$.
* You can solve a quadratic equation of the form $ax^2 + bx + c = k$, by drawing the graphs of $y = ax^2 + bx + c$ and $y = k$ and finding the value(s) of x where the graphs intersect.

Guided

1 Here is the graph of $y = x^2 - 3x - 1$.

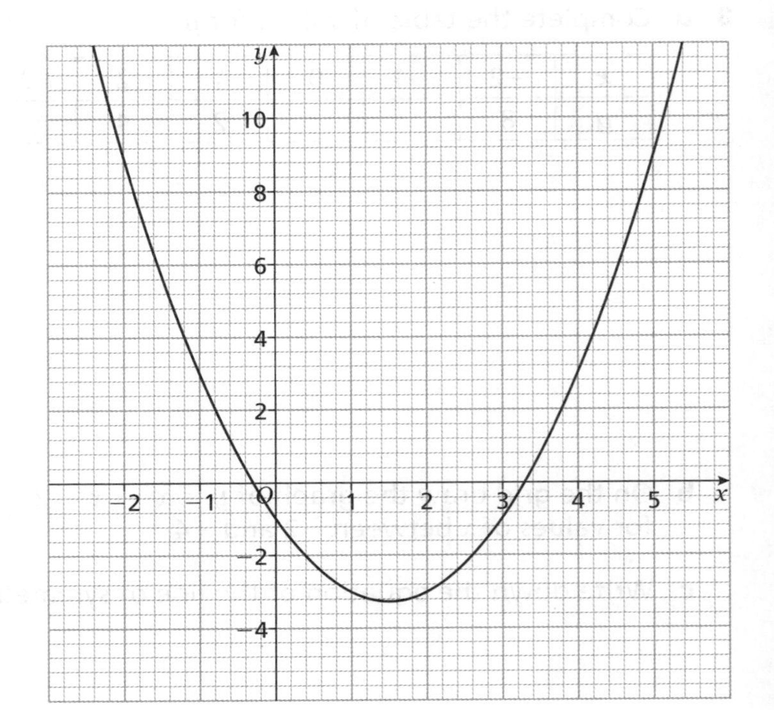

a Use the graph to find estimates for the solutions of $x^2 - 3x - 1 = 0$.

Graph is $y = x^2 - 3x - 1$:

if $x^2 - 3x - 1 = 0$ then $y = 0$.

Solutions are values of x where the graph crosses the x-axis,

i.e. $x =$ .... and $x =$ ....

b Use the graph to find estimates for the solutions of $x^2 - 3x - 1 = 6$.

Graph is $y = x^2 - 3x - 1$:

if $x^2 - 3x - 1 = 6$ then $y = 6$.

Draw in the line $y = 6$.

Solutions are values of x where the line $y = 6$ crosses the curve $y = x^2 - 3x - 1$,

i.e. $x =$ .... and $x =$ ....

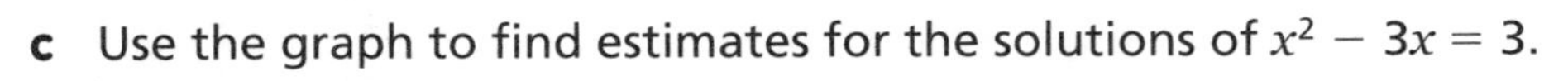

c Use the graph to find estimates for the solutions of $x^2 - 3x = 3$.

Graph is $y = x^2 - 3x - 1$:

if $x^2 - 3x = 3$

$x^2 - 3x - 1 = 3 - 1$

$x^2 - 3x - 1 = 2$

Draw in the line $y = 2$.

Solutions are values of x where the line $y = 2$ crosses the curve $y = x^2 - 3x - 1$,

i.e. $x =$ .... and $x =$ ....

Practice

2 Here is the graph of $y = x^2 + x - 5$.

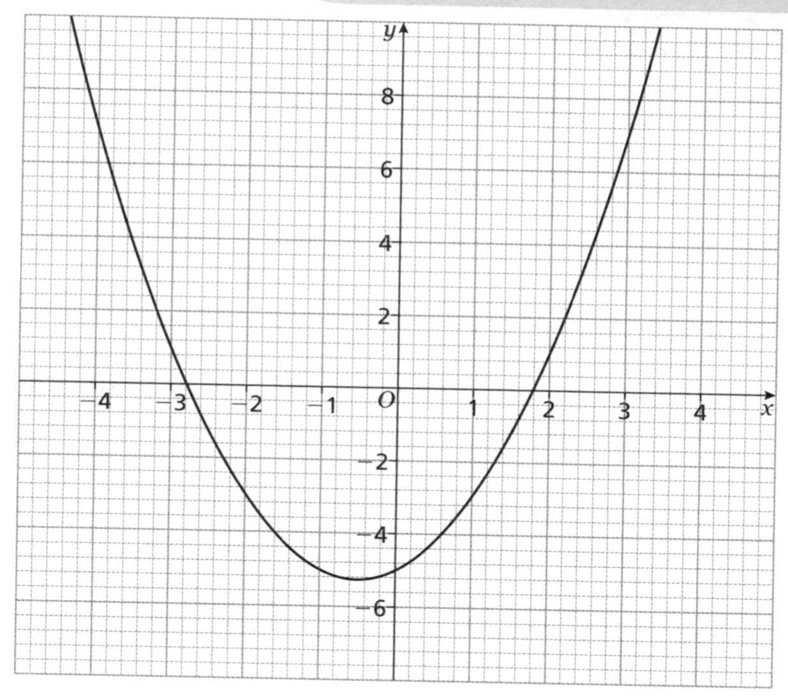

a Use the graph to find estimates for the solutions of $x^2 + x - 5 = 0$.

b Use the graph to find estimates for the solutions of $x^2 + x - 5 = -2$.

c Use the graph to find estimates for the solutions of $x^2 + x = 9$.

3 Here is the graph of $y = x^2 + 4x - 2$.

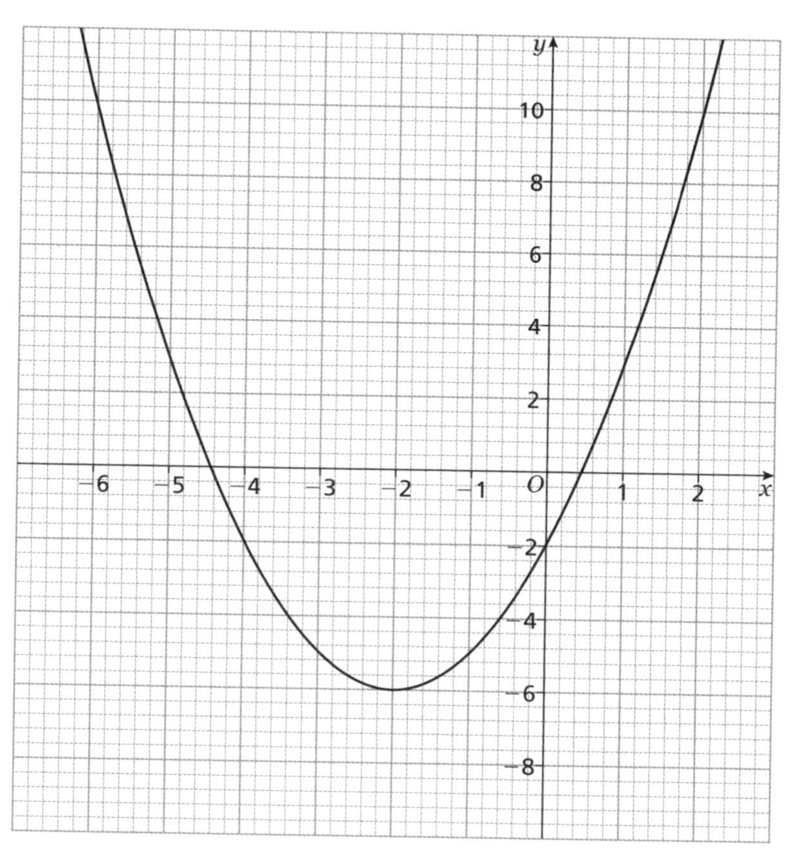

a Use the graph to find estimates for the solutions of $x^2 + 4x - 2 = 0$.

b Use the graph to find estimates for the solutions of $x^2 + 4x - 2 = -4$.

c Use the graph to find estimates for the solutions of $x^2 + 4x = 4$.

Don't forget!

- After plotting the points, join them with a smooth ....
- For graphs of functions of the form $y = ax^2 + b$, the y-axis is a .... of ....
- x^2 is always ....
- The y-intercept for a graph with equation $y = ax^2 + bx + c$ is ....
- To solve a quadratic equation $ax^2 + bx + c = 0$, draw the graph of $y = ax^2 + bx + c$ and find where the graph crosses the ....
- To solve a quadratic equation $ax^2 + bx + c = k$, draw the graphs of $y = ax^2 + bx + c$ and ...., then find the value(s) of x where the graphs ....
- To solve the equation $x^2 - 5x = 4$, draw the graphs of $y =$ .... and $y =$ .... on the same axes.

Exam-style questions

1 On the grid below, draw the graph of $y = 5x^2$ for values of x from -3 to $+3$

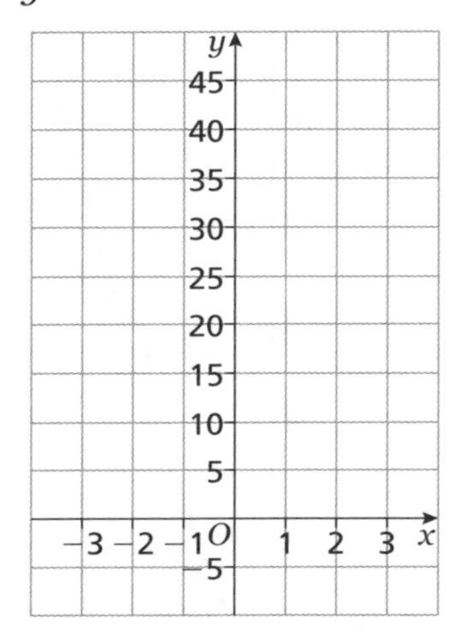

2 On the grid below, draw the graph of $y = 3x^2 - 2$ for values of x from -3 to $+3$

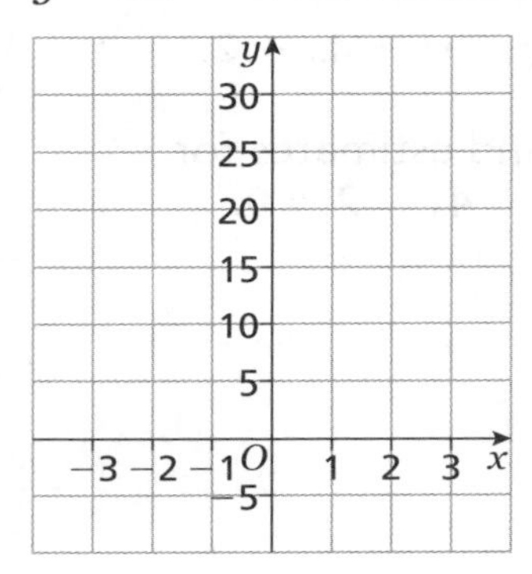

3 **a** Complete the table of values for $y = 3 - 2x^2$

x	-3	-2	-1	0	1	2	3
y	-15		1	3			

b On the grid draw the line with equation $y = 3 - 2x^2$ for values of x from -3 to $+3$

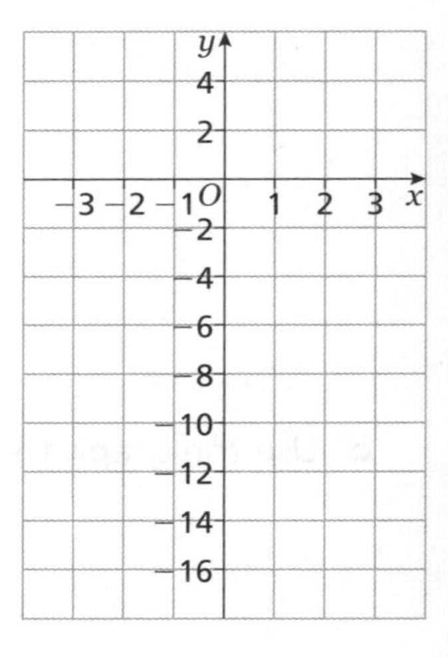

4 a Complete the table of values for $y = x^2 - 2x - 4$

x	-2	-1	0	1	2	3	4
y	4			-5	-4		

b On the grid draw the line with equation $y = x^2 - 2x - 4$ for values of x from -2 to $+4$

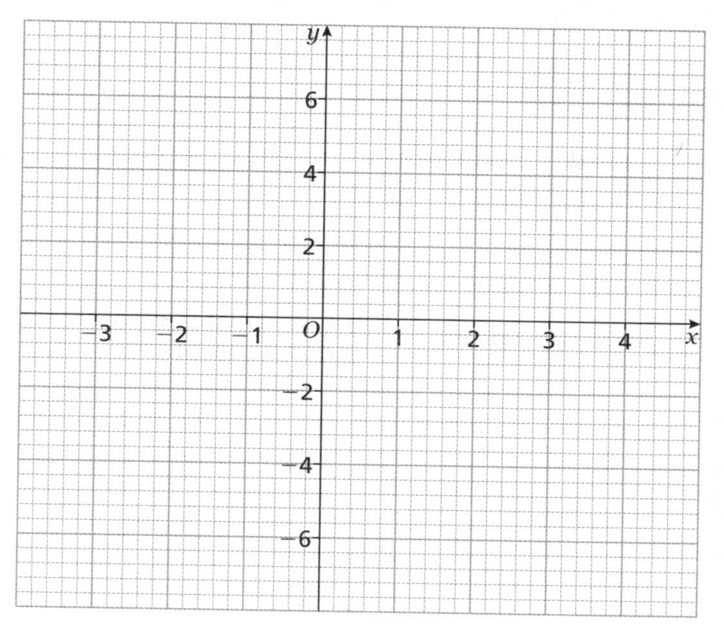

c Write down the equation of the line of symmetry of the graph.

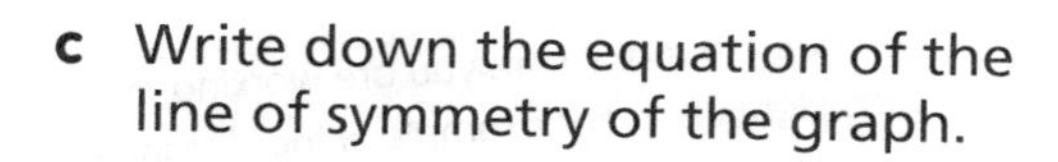

....

d Use your graph to find estimates for the solutions of $x^2 - 2x - 4 = 0$

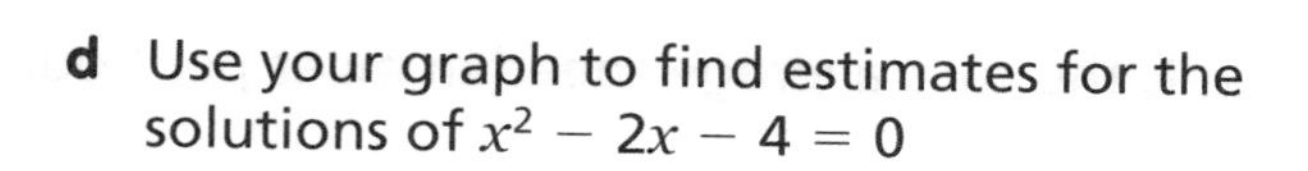

....

e Use the graph to find estimates for the solutions of $x^2 - 2x = 7$

....

5 a Complete the table of values for $y = x^2 - 6x$

x	0	1	2	3	4	5	6
y		-5	-8				

b On the grid draw the line with equation $y = x^2 - 6x$ for values of x from 0 to +6

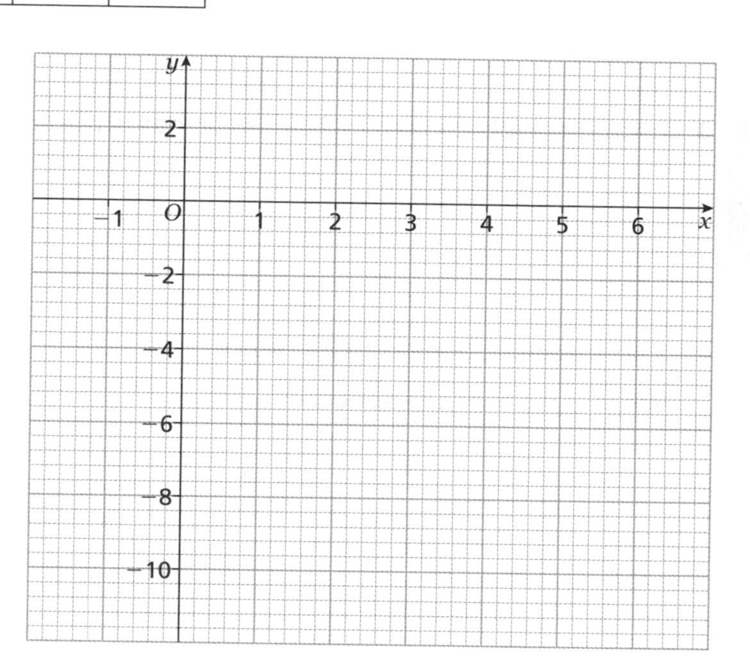

c Draw and write down the equation of the line of symmetry of the graph.

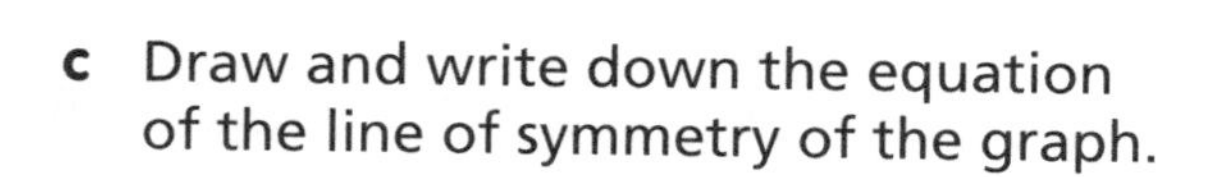

....

d Use your graph to find estimates for the solutions of $x^2 - 6x = -6$

....

GCSE LINKS
AH: 7.3 Speed;
BH: Unit 2 9.8 Speed

12.1 Speed

By the end of this section you will know how to:

* Calculate average speed given the distance covered and time taken

Key points

* **Speed** is a **compound** measure because it involves a unit of length and a unit of time.
* Common units are metres per second (m/s), km per hour (km/h) and miles per hour (mph).
* **Average speed** $= \frac{\text{total distance travelled}}{\text{total time taken}}$.

Remember this

You need to make sure you are working in consistent units. For example, if the distance is in metres and the time is in minutes, dividing would give a speed in metres per minute. To find the speed in the more common unit of m/s you need to change the minutes to seconds.

Guided

1 A car covers a distance of 120 miles in 3 hours. What is the car's average speed in miles per hour?

Average speed $= \frac{\text{total distance travelled}}{\text{total time taken}}$

$= \frac{120}{3}$

$=$ .... mph

2 A man walks 2 km in 30 minutes. What is his average speed in km/h?

Hint

Find the number of km the man walks in 1 hour.

Remember this

Sometimes you can use proportion to work out the speed. If you need to find out the speed in km/h, then you need to work out how many km are travelled in 60 minutes. For example, if in 15 minutes a car travels 20 km, then in 60 minutes it would travel $4 \times 20 = 80$ km, so the speed is 80 km/h.

1 hour = 60 min

He walks 2 km in 30 min (multiply both by 2)

= 4 km in 60 min

= 4 km in 1 hour

Answer: .... km/h

Practice

3 A train travels 260 km in 2 hours. What is the train's average speed in km/h?

.... km/h

4 Karl travels 200 miles in $2\frac{1}{2}$ hours.
What is his average speed in mph?

Hint

If a calculation involves dividing by a number such as $2\frac{1}{2}$ or 2.5, first multiply the top and bottom by 2 to get rid of the decimal or fraction.

.... mph

5 Hussein cycles 10 km in 20 minutes. What is his average speed in km/h?

Hint

$20 \times 3 = 60$ min $= 1$ hour
$15 \times 4 = 60$ min $= 1$ hour

.... km/h

Needs more practice ☐ Almost there ☐ I'm proficient! ☐

12.2 Distance–time graphs

By the end of this section you will know how to:

* Draw and interpret distance–time graphs

GCSE LINKS
AH: 15.6 Real-life graphs; **BH:** Unit 2 15.6 Real-life graphs; **16+:** 8.5 Draw and interpret distance–time graphs

Key points

* On a distance–time graph:
  - **distance** is on the vertical axis
  - **time** is on the horizontal axis
  - straight lines represent **constant speed**
  - horizontal lines represent no movement
  - the **gradient** represents the **speed**.

Remember this
As the gradient represents the speed, the **steeper** the graph the **faster** the speed.

Remember this
A distance–time graph is also called a **travel** graph.

Guided

1 Kai walks from his house to the cinema. The travel graph shows his journey.

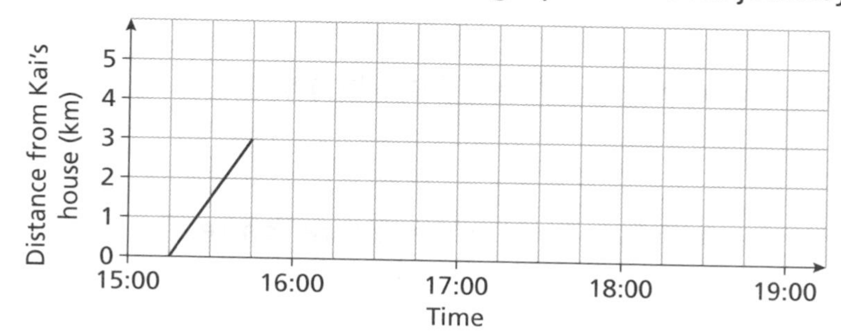

a What time does Kai leave his house?

Scale on horizontal axis: 4 squares = 1 hour = 60 min

1 square = 60 ÷ 4 = ... min

Kai leaves home at ....

b How far is the cinema from Kai's house?

The cinema is ...... km from Kai's house

c What time does Kai get to the cinema?

Kai gets to the cinema at ....

d Find Kai's average walking speed, in km/h, on the way to the cinema.

Kai walks 3 km in 30 min

= 6 km in ...... min

Answer: ...... km/h

Kai spends 2 hours at the cinema and then walks home. He arrives home at 18:30.

e Complete the travel graph.

Kai stays at the cinema for 2 hours until .... This is shown by a horizontal line on the graph.

Kai arrives home at 18:30. Join the end of the horizontal line to 18:30 on the Time axis.

f Did Kai walk faster on the way to the cinema or on the way back from the cinema? Explain your answer.

The gradient of the line is steeper on the way .... the cinema.

So Kai walked faster on the way .... the cinema.

Check: journey to cinema takes ...... min; journey back from cinema takes ...... min.

Practice

2 Matt cycles from his home to the leisure centre and back. The travel graph shows his journey.

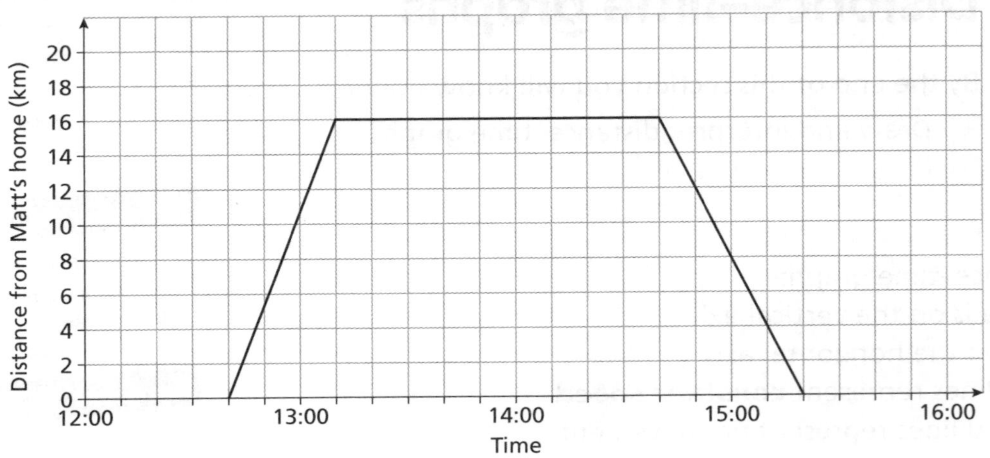

a What time does Matt leave home?

....

b How far, in km, is the leisure centre from Matt's home?

.... km

c Find Matt's speed, in km/h, on the way to the leisure centre.

.... km/h

d How long does Matt stay at the leisure centre?

....

e Find Matt's speed, in km/h, on the way back from the leisure centre.

.... km/h

3 The graph shows Becky's journey to see her friend.

a Write a story of the journey, explaining what happened during each part of it.

....

....

....

....

....

....

Distance (miles)
100
90
80
70
60
50
40
30
20
10
0
0 1 2 3 4 5 6 7 8 9 10
Time (hours)

b Work out Becky's average speed, in miles per hour, during each part of the journey.

Guided

4 Lily travels to college by bus. It takes her 10 minutes to walk 500 metres from her home to the bus stop. Then Lily waits for 15 minutes for the bus. She has a 7 km journey on the bus, which takes 10 minutes.

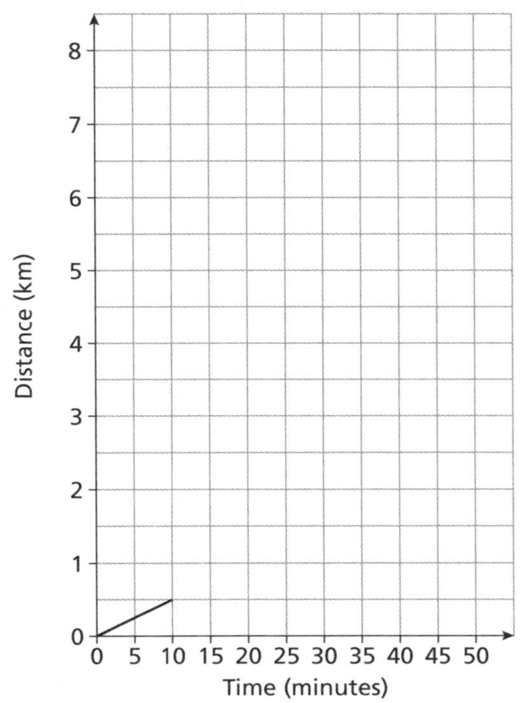

a Draw a distance–time graph of her journey.

Waiting at bus stop is horizontal line from (10, 0.5) to (25, 0.5)

Travelling on bus is line from (25, 0.5) to (......, 7.5)

b Work out the average speed of the bus in kilometres per hour.

Bus travels 7 km in 10 min

= .... km in 60 min

Average speed = ... km/h

Practice

5 Joe drives from his home to the shopping centre. The shopping centre is 20 km from Joe's home. He sets off at 2 pm and arrives at 2.30 pm. Joe spends $1\frac{1}{2}$ hours at the shopping centre and sets off home. He drives 10 km in 15 minutes then stops at a petrol station to buy some petrol. Joe spends 15 minutes at the petrol station before driving home. He arrives home at 4.45 pm.

a Draw a distance–time graph for the journey.

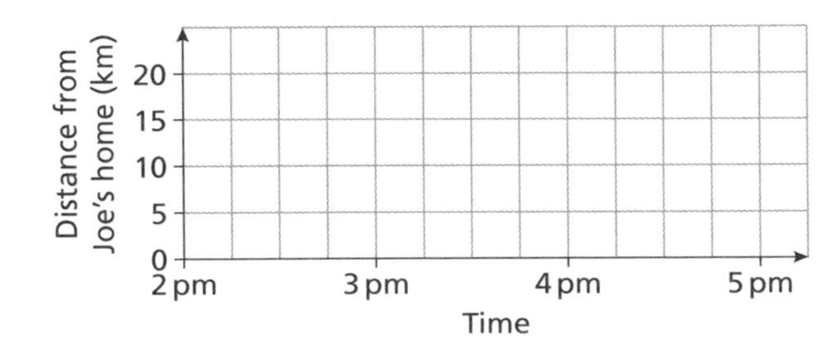

b Work out Joe's average speed, in km/h, on the way to the shopping centre.

.... km/h

Needs more practice ☐ Almost there ☐ I'm proficient! ☐

12.3 Speed–time graphs

By the end of this section you will know how to:

* Draw and interpret speed–time graphs

Key points

* On a speed–time graph:
  - **speed** is on the vertical axis
  - **time** is on the horizontal axis
  - horizontal lines represent **constant speed**
  - a **positive** gradient means the speed is **increasing**
  - a **negative** gradient means the speed is **decreasing**
  - the distance travelled is the **area** under the graph.

Remember this

If the speed of a car is increasing, then the car is **accelerating**. If the speed is decreasing, then the car is **decelerating**.

Remember this

If something 'starts from rest' it means the speed at the start, or **initial** speed, is zero.

Guided

1 A car travels at 30 m/s for 2 minutes.

a Show this information on a speed–time graph.

2 minutes = ... seconds

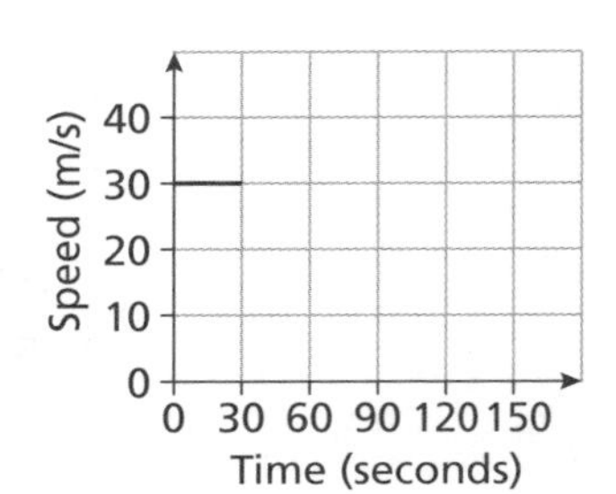

Hint

Car travels at constant speed so line is horizontal.

b What distance, in metres, does the car travel in this time?

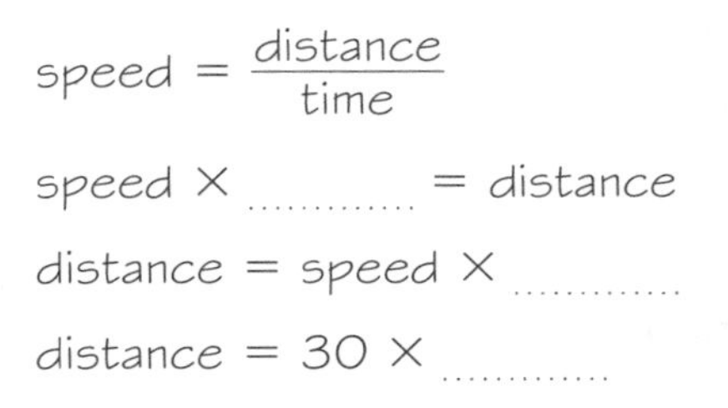

speed = $\frac{\text{distance}}{\text{time}}$

speed × .... = distance

distance = speed × ....

distance = 30 × ....

= .... m

Hint

Make distance the subject of the formula.

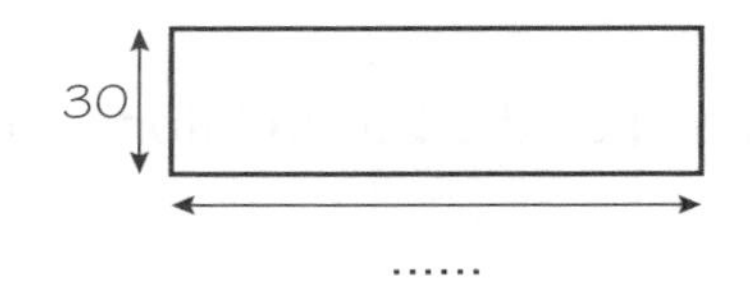

Hint

In a speed–time graph the distance travelled is the **area** under the graph. In this example it is the area of a rectangle.

2 The speed–time graph shows the journey of a scooter between two sets of traffic lights.

Speed (m/s) 0 2 4 6 8 10 12; Time (seconds) 0 5 10 15 20; regions A, B, C; stages ①, ②, ③

a Describe the scooter's journey.

① The scooter goes from 0 to 12 m/s in 5 seconds

② The scooter travels at 12 m/s for 10 seconds

③ The scooter ....

b What is the total distance, in metres, between the two sets of traffic lights?

Hint

The total distance between the two sets of traffic lights is the area under the graph.

Area of triangle **A** = $\frac{1}{2}$ × base × height

= $\frac{1}{2}$ × 5 × 12

= ...... m

Area of rectangle **B** = base × height

= ...... × 12

= .... m

Area of triangle **C** = area of triangle **A** = ... m

So total distance between traffic lights = ... + ... + ...

= .... m

3 A car remains stationary for 2 seconds and then accelerates to a speed of 8 m/s in 2 seconds. The car then travels at 8 m/s for 4 seconds and then takes 6 seconds to slow down to a halt.

a Show this information on a speed–time graph.

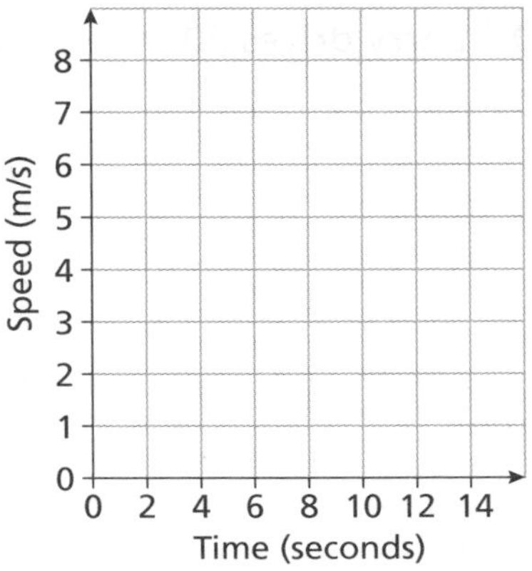

b What distance, in metres, does the car travel in this time?

Speed (m/s)
0 1 2 3 4 5 6 7 8
Time (seconds)
0 2 4 6 8 10 12 14

.... m

4 The speed–time graph shows a car's journey.

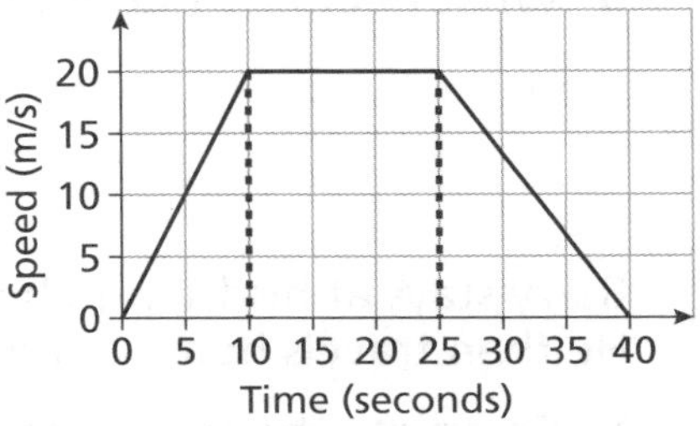

Speed (m/s)
0 5 10 15 20
Time (seconds)
0 5 10 15 20 25 30 35 40

a Describe the car's journey.

....

....

....

....

....

b What is the total distance, in metres, the car has travelled?

.... m

Don't forget!

* .... $= \dfrac{\text{total distance travelled}}{\text{total time taken}}$

* On a distance–time graph:
  - straight lines represent ....
  - .... lines represent no movement
  - the gradient represents ....
  - the steeper the graph the .... the speed.

* On a speed–time graph:
  - .... lines represent constant speed
  - a .... gradient means the speed is increasing
  - a .... gradient means the speed is decreasing
  - the area under the graph represents the .... travelled.

Exam-style questions

1 Barry drives 30 miles to his friend's house. The travel graph shows his journey.

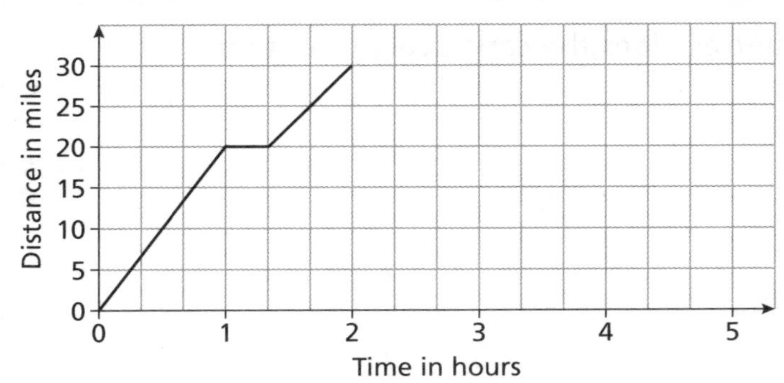

a What happens 1 hour into the journey?

....

Barry stays at his friend's house for one hour and 20 minutes.
He then travels home at an average speed of 45 mph.

b Complete the travel graph to show this information.

2 Khalid cycles to the gym which is 9 km away from his home. He leaves home at 11.15 am and arrives at the gym at 11.45 am. Khalid spends one hour and 45 minutes at the gym and then cycles home. He arrives home at 14:00

a Draw a distance–time graph of his journey.

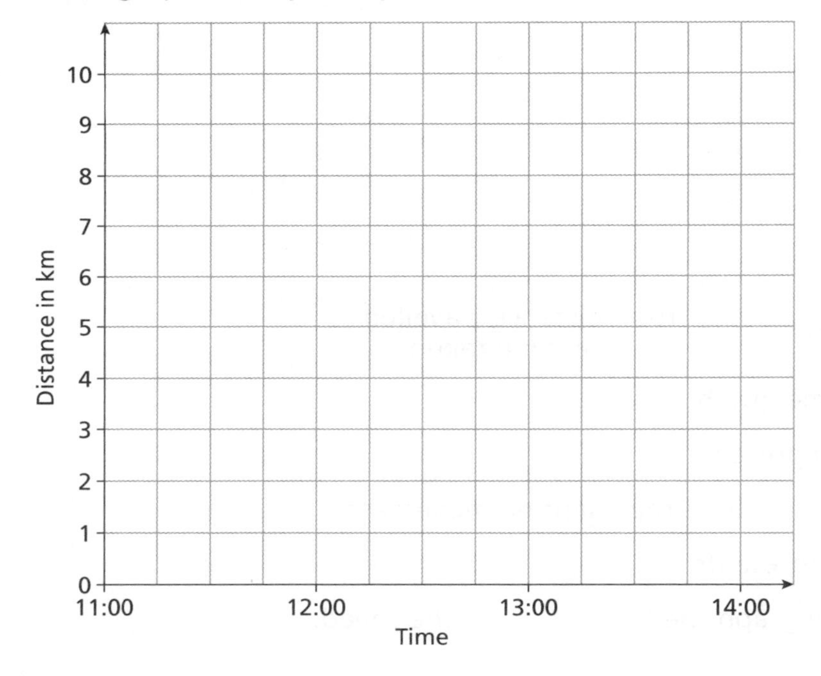

b Work out the average speed, in km/h, of Khalid's cycle ride to the gym.

.... km/h

3 A man travels upwards in a lift. The lift starts from rest and reaches a speed of 1 m/s after 2.5 seconds. The lift then travels at this speed for 5 seconds before slowing down to a halt after a further 2.5 seconds.

a Draw a speed–time graph for the lift's journey.

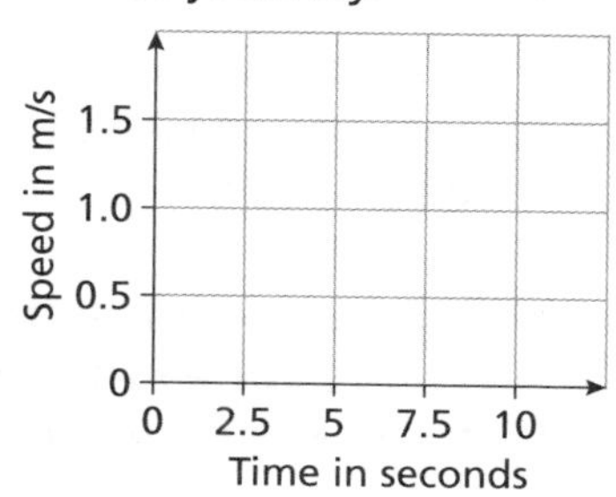

b What is the total distance, in metres, travelled by the lift?

.... m

4 The speed–time graph shows a truck's journey.

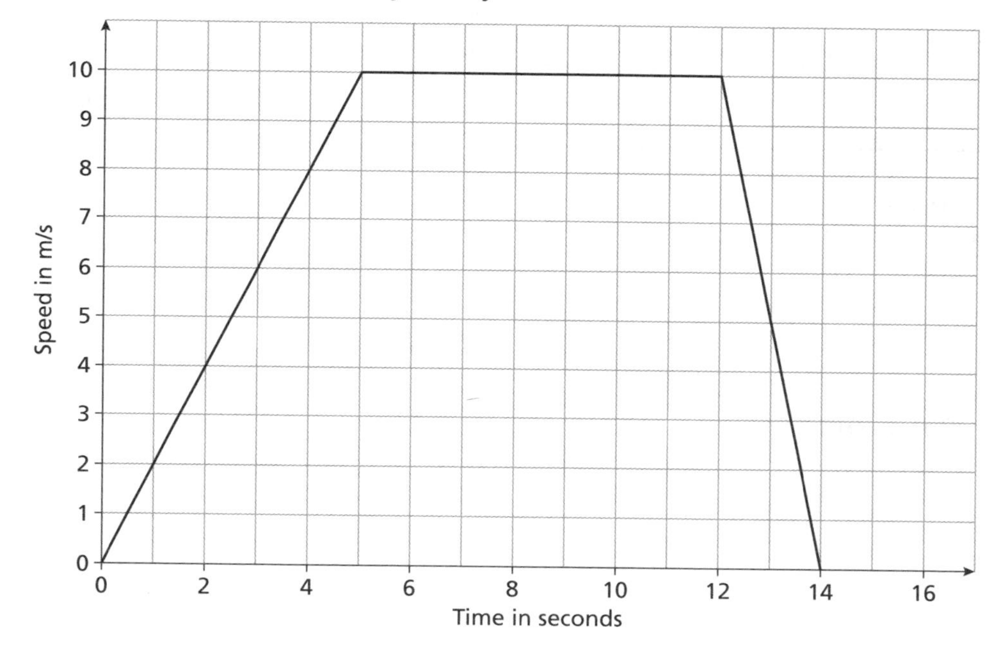

a Describe the truck's journey.

....

....

....

....

b What is the total distance, in metres, the truck has travelled?

.... m

Practice Paper

Time: 1 hour 30 minutes

Edexcel publishes Sample Assessment Material on its website. This Practice Exam Paper has been written to help you practise what you have learned and may not be representative of a real exam paper.

1 Jan sells small jars of honey for £3 and large jars of honey for £5.
On one day Jan sells s small jars of honey and l large jars of honey.
Write an expression, in terms of s and l, for the amount of money, in pounds, Jan receives for these sales.

....

(Total for Question 1 is 2 marks)

---

2 Three of the following are formulae. Tick (✓) them.

$V = IR$	☐	$a^2 - b^2 + c^2$	☐	$A = \frac{1}{2}bh$	☐
$x^2 + 5$	☐	$C = 2\pi r$	☐	$6 - 3p = 10$	☐

(Total for Question 2 is 3 marks)

---

3 **a** Simplify $b^5 \times b^3$

....

(1)

b Simplify $8p^5 \div 2p$

....

(1)

c Simplify $(2k^2)^3$

....

(2)

d Simplify $4x^2y^3 \times 3x^4y$

....

(2)

(Total for Question 3 is 6 marks)

---

4 A taxi hire company uses this formula to work out the cost of hiring a taxi.

Cost = fixed fee + cost per mile × number of miles travelled

The fixed fee is £20 and the cost per mile is £1.50

a How much does the company charge to hire a taxi for a distance of 70 miles?

£ ....

(3)

b The company charges Freddie £80 to hire a taxi. What is the distance travelled?

.... miles

(3)

(Total for Question 4 is 6 marks)

---

5 **a** Simplify $6g^2 - 7h^2 + 5 - 3g^2 - h^2 + 3$

....

(2)

b Expand $4(6s^2 - 7)$

....

(2)

c Expand $3w(2w^2 - 5w - 1)$

....

(2)

d Expand and simplify $2v(5v - 4) - 3v(4v - 7)$

....

(3)

(Total for Question 5 is 9 marks)

---

6 $y = mx + c$

a Find the value of y when $m = 2$, $x = 3$ and $c = -5$

....

(2)

b Find the value of c when $y = 2$, $m = -1$ and $x = 4$

....

(3)

c Make x the subject of the formula.

$x =$ ....

(3)

(Total for Question 6 is 8 marks)

---

7 The nth term of a sequence is $4n - 3$

a What is the tenth term of the sequence?

....

(2)

b Is 200 a term of the sequence?
Explain your answer.

....

....

(2)

(Total for Question 7 is 4 marks)

---

8 **a** Factorise $4xy + 8x$

....

(2)

b Factorise $2a^4b - 4a^3b^2 + 6a^2b$

....

(2)

(Total for Question 8 is 4 marks)

---

9 **a** Solve $\frac{b-3}{4} = 5$

....

(2)

b Solve $3w + 2 = 4 - 5w$

....

(2)

c Solve $2(v + 4) = 7$

....

(3)

(Total for Question 9 is 7 marks)

---

10 Sketch the graph of $y = x^2 - 16$
You must label the points where the graph crosses the x and y axes.

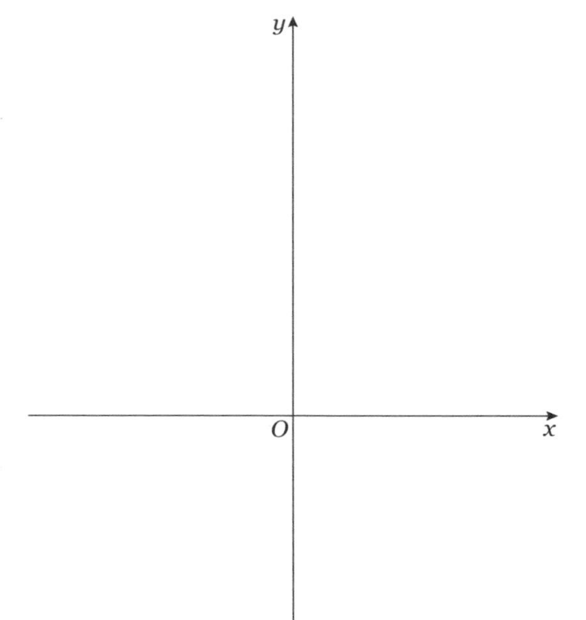

....

(Total for Question 10 is 4 marks)

11 a List the integers that satisfy the inequality $-4 < n \leqslant 2$

....

(2)

b Find the smallest possible integer value of x that satisfies the inequality $3x - 5 \geqslant 9$

....

(4)

(Total for Question 11 is 6 marks)

---

12 a Find the gradient of the straight line joining the points A (−2, 7) and B (3, −3).

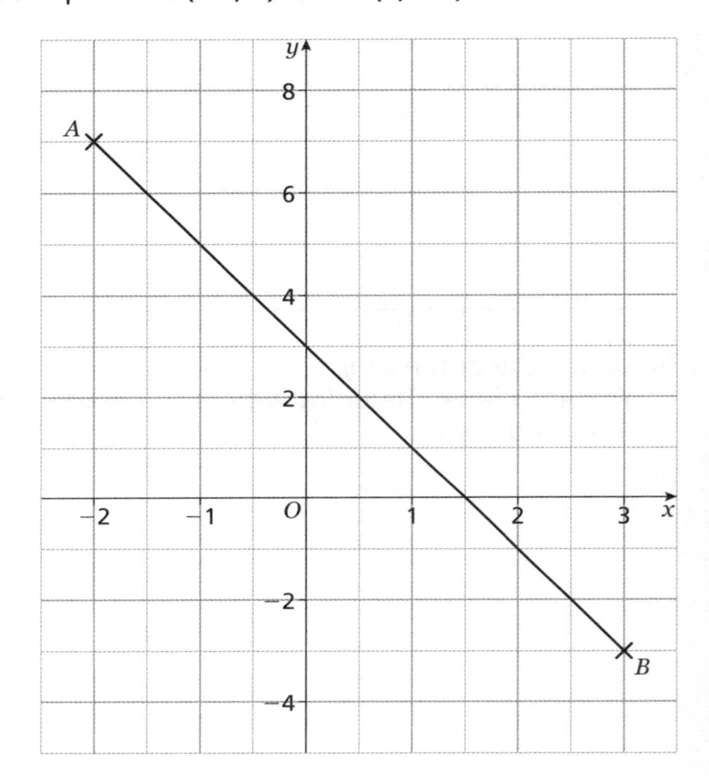

....

(2)

b Find an equation of the straight line joining the points A and B.

....

(2)

(Total for Question 12 is 4 marks)

---

13 Jackie compares the cost of joining three different gyms. The graph shows the monthly cost of membership at Gym A, Gym B and Gym C.

Gym A	£20 joining fee plus £15 per month
Gym B	No joining fee, £20 per month
Gym C	£35 joining fee plus £12 per month

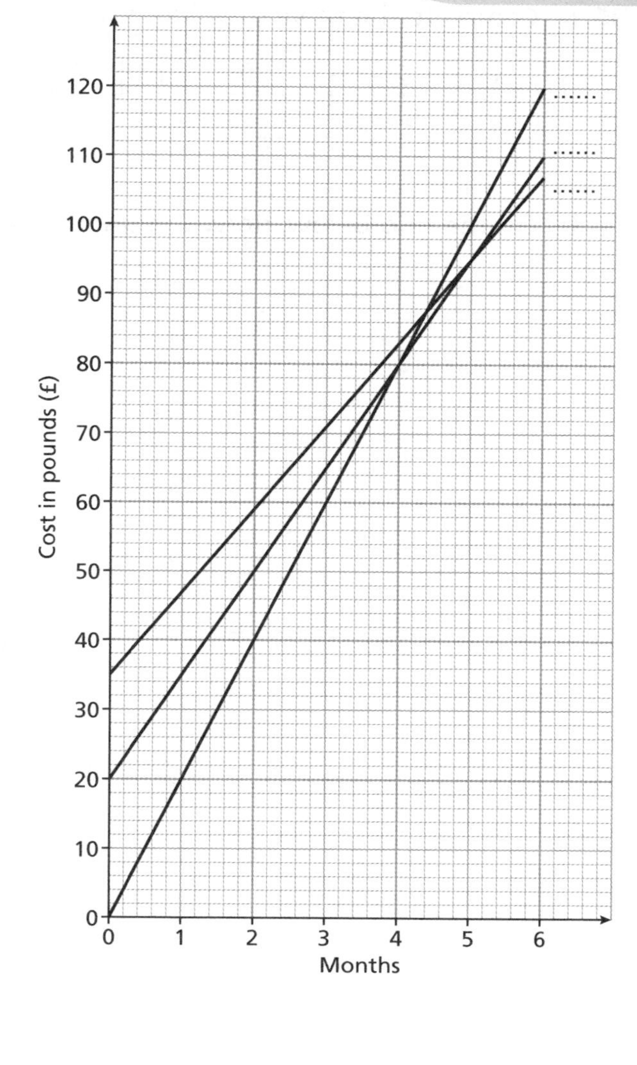

a Label each line on the graph with the gym it represents. (2)

b Give Jackie some advice, based on costs, about how to choose which gym she should join. You must give the reasons for your answer.

....

....

....

(2)

(Total for Question 13 is 4 marks)

---

14 The speed–time graph shows the motion of a train over a 20-second period.

Speed in m/s: 0, 10, 20, 30, 40, 50, 60

Time in seconds: 0, 5, 10, 15, 20

a Describe the motion of the train.

....

....

....

(2)

b Find the total distance travelled by the train in the 20-second period.

.... metres

(3)

(Total for Question 14 is 5 marks)

15 a Complete the table of values for $y = x^2 - 3x - 5$

x	−2	−1	0	1	2	3	4	5
y	5		−5	−7			−1	

(2)

b On the grid below, draw the graph of $y = x^2 - 3x - 5$ for values of x from −2 to 5

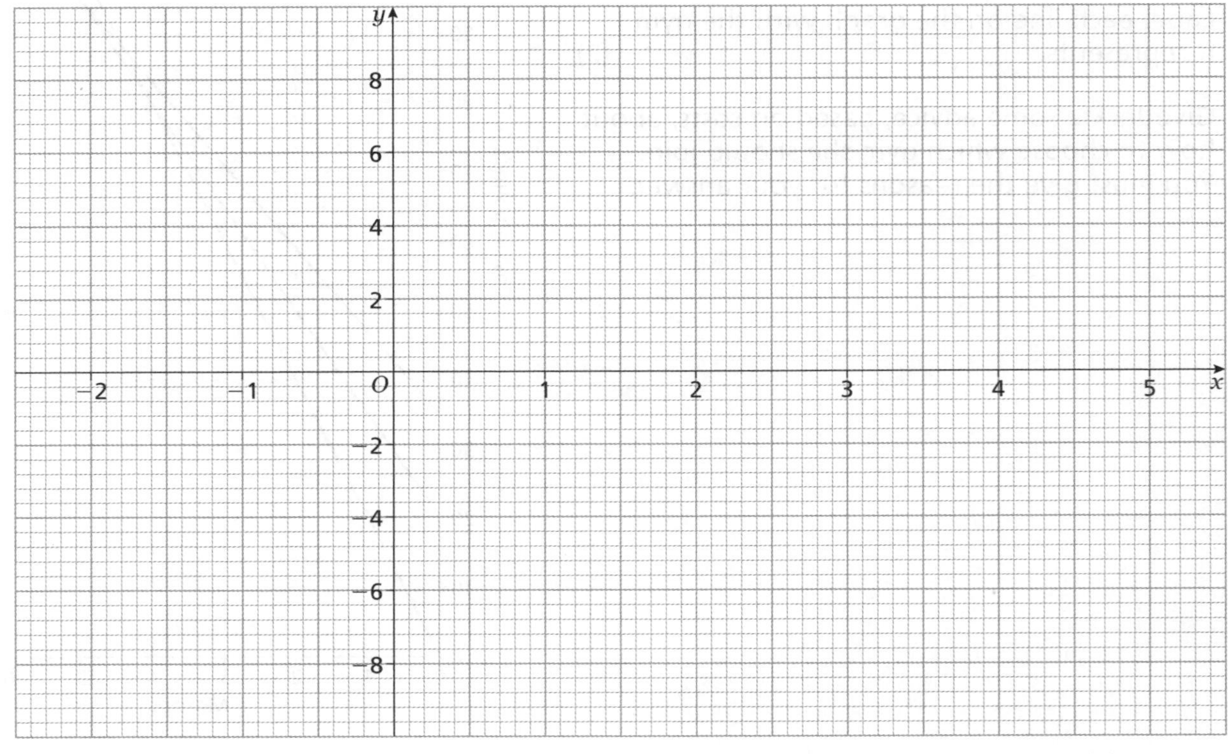

(2)

c Use your graph to find estimates for the solutions of $x^2 - 3x - 5 = 0$

....

(2)

d Use your graph to find estimates for the solutions of $x^2 - 3x = 2$

....

(2)

(Total for Question 15 is 8 marks)

TOTAL FOR PAPER IS 80 MARKS

Answers

1 Roles of symbols

1.1 Using letters to represent numbers

1 a $4a$ b $3x$ c $5c$ d $3b$
2 a $7a$ b $4b$ c $-c$
3 a $7k$ b $8y$ c $11x$ d $-2y$
e $7k$ f $-3m$
4 a $12a$ b b c $3c$ d $-d$

1.2 Equations, formulae and expressions

1 a and c
2 a $A = \frac{1}{2}bh$ b $P = 2(l + w)$ c $V = lwh$
3 a equation b expression c formula
4 a equation b formula c equation d expression
5 a equation b expression c formula d formula

1.3 Representing situations in real life

1 $b + 4$ 2 $s - 2$ 3 $d + 6$
4 $a - 3$ 5 $y - 3$ 6 $x + y$
7 $3m$ 8 $6p, 12q, 6p + 12q$
9 $4x + 8y$ 10 $4e + 9f$
11 $5g + 8h$ 12 $\frac{90c + 75t}{100}$ or $\frac{18c + 15t}{20}$

Don't forget!

* $P = 2l + 2w \rightarrow$ formula; $8 + 4a - 5b \rightarrow$ expression; $3x - 7 = 2 \rightarrow$ equation
* terms, expression
* $y + 2$; $y - 1$
* $3x$; $-4y$

Exam-style questions

1 $9a$ 2 $-6y$
3 $M = D \times V$; $a^2 + b^2 = c^2$; $A = l \times w$
4 $12k^2 + 7 = 33$; $14 = 5x - 1$; $4x - 16 = 0$
5 $b + w$ 6 $c + 5$ 7 $p - 3$
8 $\frac{k}{3}$ or $\frac{1}{3}k$ 9 $3p + 10q$ 10 $6d + 4c$
11 $\frac{80w + 95c}{100}$ or $\frac{16w + 19c}{20}$

2 Algebraic manipulation

2.1 Collecting like terms

1 a $9x - 14$ b $2a - 3b$
c $9 + 2 - 8c - 4c + 5d + 3d = 11 - 12c + 8d$
2 a $3y - 2x - 7$ b $5p + 3r + 5$
c $2b - 3a - 2$ d $4t - s - 11$
3 a $x - 3x^2$ b $2xy - 3$
c $3x^2 + 6x^2 + 5y^2 - 7y^2 - 9xy - 2xy = 9x^2 - 2y^2 - 11xy$
d $7y^2 - 4y^2 + 2x^2 - 5x^2 + xy - 3xy = 3y^2 - 3x^2 - 2xy$
4 a $7k^2 + k - 1$ b $5vw + 7w + 15$
c $6h^2 - 3g^2 - 3$ d $3a^2 - 11ab - b^2 + 8$
5 a $4e^3 + 4e^2 + 8e$ b $12c^2 - 8c + 8$
6 $6x + 11y$

2.2 Multiplication with brackets

1 a $5y + 40$
b $3 \times 2x + 3 \times -1 = 6x - 3$
c $4 \times 6a + 4 \times -3b = 24a - 12b$
d $-2 \times 5pq - 2 \times 4q^2 = -10pq - 8q^2$
e $-3 \times 4w^2 - 3 \times -7v^2 = -12w^2 + 21v^2 = 21v^2 - 12w^2$
f $-1 \times 3xy - 1 \times -2y^2 = -3xy + 2y^2 = 2y^2 - 3xy$
2 a $21x + 35 + 12x - 48 = 33x - 13$
b $40p - 16 - 12p - 27 = 28p - 43$
c $27s + 9 - 30s + 50 = -3s + 59 = 59 - 3s$
d $8x - 6 - 3x - 5 = 5x - 11$
3 a $14 - 6b^2$ b $15pq - 5p^2$
c $12y^2 - 8x^2$ d $21a^2 - 15b^2$
4 a $22k + 27$ b $-3y - 27$ c $13x - 17$ d $22a - 25$
5 a $6a^2 - 14a$
b $3r \times 4r^2 + 3r \times -8 = 12r^3 - 24r$
c $-10s^3 + 45s^2 - 50s = 45s^2 - 10s^3 - 50s$
d $-6x^3 + 21x^2 - 15x = 21x^2 - 6x^3 - 15x$
6 a $12x^2 + 24x$ b $20k^3 - 48k$
c $10h - 12h^3 - 22h^2$ d $21s^2 - 21s^3 - 6s$
7 a $-y^2 - 4$ b $5x^2 - 11x$ c $2p - 7p^2$ d $6b^2$
8 $y - 4$
9 a $-1 - 2m$ b $5p^3 + 12p^2 + 27p$
10 $7x(3x - 5) = 21x^2 - 35x$

2.3 Factorising

1 a $5(2x + 3)$; Check: $5(2x + 3) = 10x + 15$
b $2(1 - 6a)$; Check: $2(1 - 6a) = 2 - 12a$
c $y(y - 3)$; Check: $y(y - 3) = y^2 - 3y$
2 a $3k(k - 9)$ b $7b(2 + 3b)$ c $8xy(y - 2x)$
3 a $4(2b - 3)$ b $3(1 + 3p)$ c $h(6 - 5h)$
4 a $2x(y - 4x)$ b $6g(2g + 3)$ c $5ab(2 - a + 3b)$
5 a $3x^2(2x - 3)$ b $2b(2a + 4c - d)$
c $4pq(3q - 4p + 1)$

2.4 Laws of indices

1 a y^6 b x^2 c 1
2 a a^6 b b^{12} c $p^{4 \times 5} = p^{20}$
3 a x^5 b y^{12} c p^2 d b
e r^{24} f m^{15}
4 a a^{13} b k^{11} c m^6 d b^7
e k^{12} f r^{21}
5 a $6 \times c^9 = 6c^9$ b $2 \times x^3 = 2x^3$
6 a $24d^{12}$ b $4y^4$ c $15k^{13}$ d $3b^7$
7 a $28 \times a^7 \times b^{11} = 28a^7b^{11}$
b $3 \times 12 \times c^2 \times c^7 \times d^3 \times d^8 = 36 \times c^9 \times d^{11} = 36c^9d^{11}$
c $16 \times a^{12} = 16a^{12}$
d $3^2 \times (b^5)^2 = 9 \times b^{10} = 9b^{10}$
8 a $30x^8y^5$ b $32p^8q^8$ c $32x^{20}$ d $81y^{24}$
9 a $30p^4q^4$ b $256x^8y^{12}$ c $\frac{3a^2}{b}$

Don't forget!

* $4a$; a^4; $7k - 7j$; $5r^2 - 3r - 12$
* $15t - 21$; $3s(6 + s)$; x^{m+n}; x^{m-n}; x^{mn}; x^0; x^1; p^9; p^4; b^{15}; $15a^5b^9$; $2y^2$

Exam-style questions

1 a $3k + 3k^2 - 2$ b x^9 c s^6 d g^{21}
2 a $20r - 24s$ b $6g^3 - 21g^2 + 24g$
3 a $3h - 29$ b $10x^2 - 22x$
4 a $3(2b - 3a + 1)$ b $2x(2x^2 + 6x - 3)$
c $5xy(x - 2y + 3)$
5 a $21a^5$ b $7p^3$ c $20v^7w^7$ d $81g^{16}$
6 a $3h^2 - 2h - 3$ b $15g^9$ c $6k^3$ d $32m^{15}$
7 a $6x^2y - 2xy^2$ b $-14k^2 - 40k$
8 a $x(3 - 5x + 7x^2)$ b $2vw(w - 2v + 4vw)$
9 a $n^7 - n^5$ b $p^7 + p^6 - p^5$

3 Formulae

3.1 Using word formulae

1 a 12 cm b $4 \times 5 = 20$ cm c $40 \div 4 = 10$ cm
2 a £48 b $£11 \times 5 = £55$ c $£60 \div 4 = £15$
3 a 40 km/h b 200 km c 2 hours
4 a 75 min b 5 pounds
5 24 cm²
6 £120
7 £21

3.2 Using algebraic formulae

1 a $s = 10$ **b** $s = 5 \times -4 = -20$ **c** $t = \frac{15}{5} = 3$

2 a $y = 13$

b $y = 3 \times -2 - 5 = -11$

c $y = 3 \times \frac{1}{3} - 5 = -4$

d $13 + 5 = 3x$
$18 = 3x$
$18 \div 3 = x$
$x = 6$

3 a $T = 320$ **b** $T = -220$ **c** $T = 35$ **d** $W = 1$

4 $V = 24$

5 $s = 5$

6 a $v = 2$ **b** $v = 45$

7 a $s = 8$ **b** $s = 16$

8 a $s = 8$ **b** $s = -6$

9 $v = +4$ or $v = -4$

10 $u = +3$ or $u = -3$

3.3 Changing the subject of a formula

1 Flowchart method: $x = \frac{y+5}{4}$ or $x = \frac{1}{4}(y + 5)$

Algebraic method: $\frac{y+5}{4} = x$

$x = \frac{y+5}{4}$

2 $W = \frac{T-20}{30}$ **3** $Q = 7R$

4 $k = 4(P + 5) = 4P + 20$ **5** $s = \frac{3}{V}$

6 $R = 5(L - 2) = 5L - 10$ **7** $t = \frac{v-u}{a}$

8 $d = st$ **9** $w = \frac{V}{lh}$

10 $l = \frac{P}{2} - w$ or $l = \frac{P-2w}{2}$ **11** $T = \frac{100I}{PR}$

12 $V = \frac{M}{D}$ **13** $a = \frac{P-3b^2}{2}$

14 $r = \sqrt{\frac{A}{\pi}}$

Don't forget!

* formula; words; symbols
* 15 cm
* $P = 10$
* $V = 5$
* $V = \frac{W}{A}$; $W = PT$; $A = \frac{F}{P}$; $x = \frac{P-3y}{2}$

Exam-style questions

1 a £90 **b** 5 hours

2 £15.50

3 a $P = 25$ **b** $x = 7$ **c** $x = \frac{P+5}{3}$

4 a $W = 50$ **b** $W = 6$

5 a $A = 6$ **b** $h = \frac{2A}{b}$

6 a $P = 13$ **b** $a = 4$ **c** $b = \frac{P-2a}{3}$

7 a $v = 15$ **b** $v = 24$

8 $m = \frac{2E}{v^2}$

9 $c = 2s - a - b$

10 $b = \frac{s-6a}{4}$

11 $h = \frac{2A}{a+b}$

4 Linear equations

4.1 Solving equations with one operation

1 a $x = 3$ **b** $y - 7 + 7 = 3 + 7; y = 10$

c $7 = z$ or $z = 7$ **d** $p = -6$

2 a $s = 5$ **b** $t = 10$ **c** $k = -2$

d $p = 19$ **e** $w = 0$ **f** $v = 21$

3 a $x = 5$ **b** $y = 12$ **c** $\frac{25k}{25} = \frac{5}{25}, k = \frac{1}{5}$

4 a $k = 5$ **b** $g = 28$ **c** $w = 9$

d $v = 18$ **e** $s = 4$ **f** $t = 40$

g $c = \frac{1}{3}$ **h** $x = \frac{1}{10}$ **i** $y = 24$

5 $x = 18$

4.2 Solving equations with two operations

1 a $4a = 8$
$\frac{4 \times a}{4} = \frac{8}{4}$
$a = 2$

b $3x = 15$
$\frac{3 \times x}{3} = \frac{15}{3}$
$x = 5$

c $\frac{1}{7}s = 1$
$\frac{s \times 7}{7} = 1 \times 7$
$s = 7$

2 a $a = 6$ **b** $b = 9$ **c** $k = 36$ **d** $m = 9$

e $a = 3$ **f** $w = 60$

3 a $5s = 14$
$\frac{5 \times s}{5} = \frac{14}{5}$
$s = \frac{14}{5}$ (or $2\frac{4}{5}$ or 2.8)

b $3k = -6$
$\frac{3 \times k}{3} = \frac{-6}{3}$
$k = -2$

4 a $v = \frac{1}{3}$ **b** $p = -1$ **c** $w = -5$ **d** $k = \frac{13}{8}$

e $x = -\frac{1}{3}$ **f** $y = -\frac{2}{5}$

5 a $6x + 120 = 360$ **b** $x = 40°$

4.3 Solving equations with brackets

1 a $4x - 4 + 4 = 5 + 4$
$4x = 9$
$\frac{4x}{4} = \frac{9}{4}$
$x = \frac{9}{4}$

b $8w + 26 = 20$
$8w + 26 - 26 = 20 - 26$
$8w = -6$
$\frac{8w}{8} = \frac{-6}{8}$
$w = -\frac{3}{4}$

c $15k - 10 = 2$
$15k - 10 + 10 = 2 + 10$
$15k = 12$
$\frac{15k}{15} = \frac{12}{15}$
$k = \frac{4}{5}$

d $6p + 3 - 3 = 5 - 3$
$6p = 2$
$\frac{6p}{6} = \frac{2}{6}$
$p = \frac{1}{3}$

2 a $y = 7.5$ **b** $p = -\frac{10}{3}$ **c** $r = 4$ **d** $k = -\frac{1}{5}$

e $w = \frac{11}{12}$ **f** $x = -8$ **g** $y = 3$ **h** $x = -\frac{5}{3}$

i $y = -1$

3 $(w - 3) = 8$
$w - 3 = 8$
$w - 3 + 3 = 8 + 3$
$w = 11$

4 a $x = 16$ **b** $y = 14$ **c** $k = 5.5$

5 a $p = \frac{4}{3}$ **b** $w = 6$

4.4 Solving equations where the unknown appears on both sides

1 a Method 1: $2a = 10, a = 5$ Method 2: $2a = 10, a = 5$

b Method 1: $4b = b - 9$
$4b - b = b - 9 - b$
$3b = -9$
$b = -3$

Method 2: $3b = -9$
$b = -3$

c Method 1: $3c + 12 = 6c$
$3c + 12 - 3c = 6c - 3c$
$12 = 3c$
$3c = 12$
$c = 4$

Method 2: $3c = 12$
$c = 4$

2 **a** $x = 5$ **b** $k = -4$ **c** $v = 2.5$ **d** $s = 15$
e $h = -1$ **f** $t = \frac{12}{5}$

3 $x = -\frac{11}{6}$

4.5 Solving equations with negative coefficients

1 **a** Method 1: $-4 = 2x$
$2x = -4$
$x = -2$
Method 2: $-4 = 2x$
$2x = -4$
$x = -2$

b Method 1: $-3y = 3 - 5y$
$-3y + 5y = 3 - 5y + 5y$
$2y = 3$
$y = \frac{3}{2}$
Method 2: $6 = 9 - 2y$
$6 - 9 = 9 - 2y - 9$
$-3 = -2y$
$2y = 3$
$y = \frac{3}{2}$

2 **a** $p = 1$ **b** $x = 3$ **c** $y = -4$ **d** $k = \frac{3}{10}$
e $w = -1$ **f** $s = -\frac{5}{2}$

3 $x = -\frac{2}{11}$

Don't forget!

* powers
* left-hand side, right-hand side (or vice versa)
* balance method
* + or −
* brackets, brackets
* more of them
* coefficient
* $y = 5$; $x = 6$; $w = 18$; $v = 4$

Exam-style questions

1 **a** $s = 4$ **b** $k = 3$ **c** $p = 2$ **d** $x = -\frac{5}{4}$

2 **a** $w = -6$ **b** $w = 2$ **c** $v = \frac{7}{3}$ **d** $y = 19$

3 **a** $d = 10$ **b** $t = \frac{2}{5}$ **c** $r = \frac{11}{8}$ **d** $x = 16.5$

4 **a** $x = 4$ **b** $w = 3$ **c** $k = -\frac{4}{9}$ **d** $x = \frac{16}{7}$

5 Linear inequalities

5.1 Introducing inequalities

1 **a** $9 > 5$ **b** $3 < 7$ **c** $2 > -3$

2 **a** $0 < 2$ **b** $7 > -5$ **c** $9.2 < 11.1$

3 **a** true **b** false **c** true

5.2 Representing inequalities on a number line

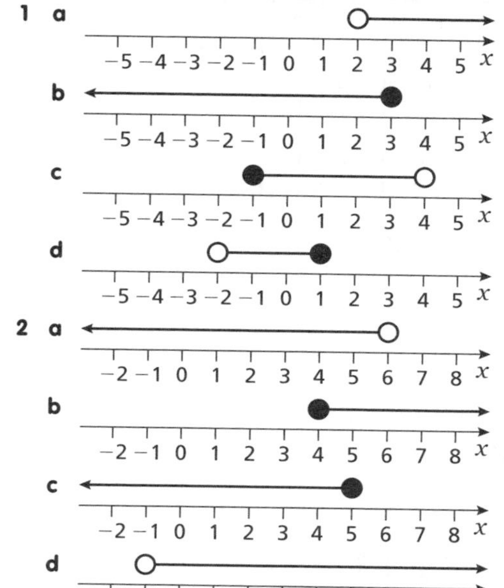

3 **a**

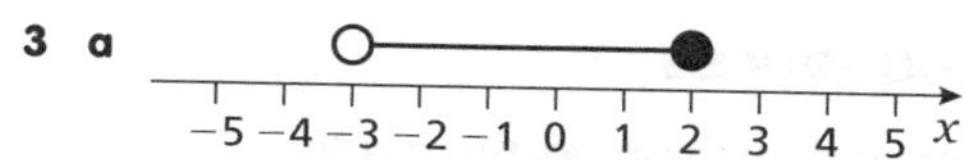

b

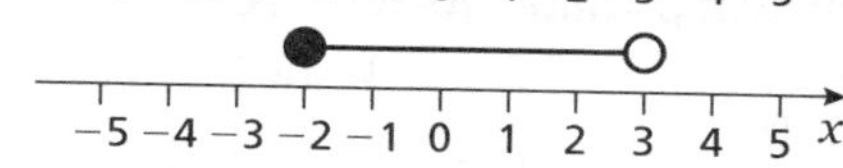

c

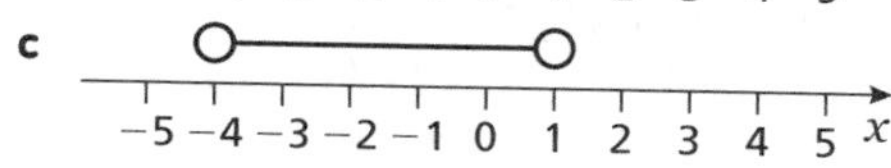

d

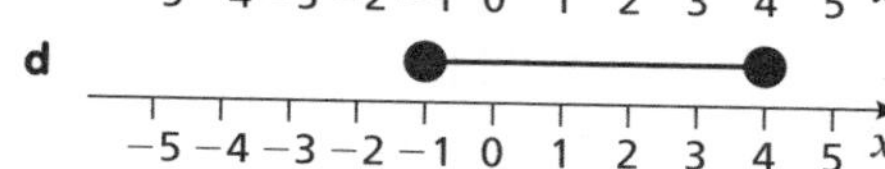

4 **a** $x < 2$ **b** $x \geqslant -1$ **c** $-4 \leqslant x < 3$
d $-1 < x \leqslant 5$

5 **a** $x < 6$ **b** $x \geqslant 4$ **c** $x > 3$

6 **a** $-3 \leqslant x < 2$ **b** $-5 < x < 0$ **c** $-2 \leqslant x \leqslant 5$

5.3 Solving linear inequalities

1 **a** $x < 12$
b $2x \geqslant 8$
$x \geqslant 4$
c $5x < 18$
$x < \frac{18}{5}$

2 **a** $x < -4$
b $6 < 2x$
$2x > 6$
$x > 3$
c $\frac{-4x}{-4} \geqslant \frac{-3}{-4}$
$x \geqslant \frac{3}{4}$

3 **a** $x > 4$ **b** $x \leqslant 2$ **c** $x \leqslant -1$
d $x > -\frac{7}{2}$ **e** $x \geqslant 10$ **f** $x < -15$

4 **a** $x < -20$ **b** $x \leqslant 3.5$ **c** $x < 4$

5 **a** $t < \frac{5}{2}$ **b** $n \geqslant \frac{7}{5}$

5.4 Finding integer solutions to inequalities

1 **a** $-3, -2, -1, 0, 1, 2, 3$
b $-1, 0, 1, 2, 3$
c $1, 2, 3, 4, 5, 6$

2 $2x > 7$
$x > 3.5$
smallest integer value = 4

3 **a** $-5, -4, -3, -2, -1, 0, 1, 2, 3, 4$
b $-4, -3, -2, -1, 0$
c $3, 4, 5, 6, 7$
d $-3, -2, -1, 0, 1, 2, 3, 4, 5$

4 $x \leqslant \frac{11}{3}$ or $3\frac{2}{3}$ so $x = 3$

5 $-2, -1, 0, 1, 2, 3$

Don't forget!

* greater than; greater than or equal to; less than; less than or equal to
* is not; is
* whole number

Exam-style questions

1 **a** $-3 < x \leqslant 2$
b
c $-3, -2, -1, 0, 1, 2, 3$
d $t \leqslant 3.5$

2 **a** $-2 \leqslant y < 3$
b
c $-5, -4, -3, -2, -1, 0$
d $x > \frac{20}{3}$

3 **a** $-4 < x \leqslant 3$
b

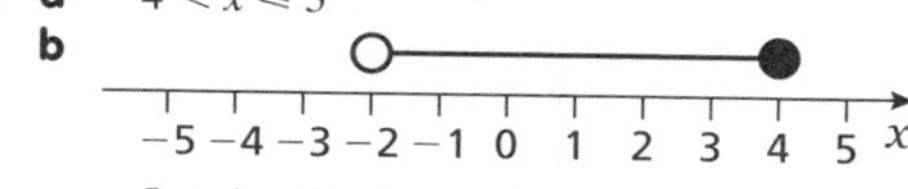

c $-5, -4, -3, -2, -1, 0, 1, 2$
d $x < -1$

4 $x < 3$

5 $x > 6.5$ so $x = 7$

6 Number sequences

6.1 Term-to-term and position-to-term rules

1 a $23 + 5 = 28$; 23, 28 **b** Add 5 to each term
c $3 + (9 \times 5) = 48$
2 a 64, 128 **b** Multiply by 2 **c** 256
3 a 46, 54 **b** Add 8 **c** 62
4 1 000 000
5 $10^3 = 1000$
6 $3^6 = 729$
7 $4^5 = 1024$
8 a 96 **b** 84
c 7 is not a multiple of 4
9 $12^2 - 1 = 143$

6.2 The nth term of an arithmetic sequence

1 a $4n + 1$ **b** $200 + 1 = 201$
2 a $3n + 2$ **b** $3 \times 100 + 2 = 302$
3 a $-2n + 114$ **b** $-2 \times 50 + 114 = 14$
4 a $-5n + 100$ **b** $-5 \times 20 + 100 = 0$
5 a $n = 1$: -4; $n = 2$: -1; $n = 3$: $3 \times 3 - 7 = 9 - 7 = 2$;
$n = 4$: $3 \times 4 - 7 = 12 - 7 = 5$;
first four terms are: $-4, -1, 2, 5$
b $3n = 72$
$n = 24$
the 24th term is 65
6 a 9, 14, 19, 24 **b** $5n + 4 = 59$, 11th term
7 a 48, 44, 40, 36 **b** 13th term
8 a $3 \times 5^2 = 75$
b $3n^2 = 300$, $n^2 = 100$, $n = 10$; yes it is the 10th term

Don't forget!

* rule; terms; term-to-term; position-to-term rule; (difference between terms) $\times\, n$ + zero term
* $4n + 6 \rightarrow 10$; $3n - 2 \rightarrow 1$; $5 - 2n \rightarrow 3$; $8 - 3n \rightarrow 5$

Exam-style questions

1 a 26, 31 **b** $5n + 1$ **c** 251
2 a $-3, -1$
b If $2n - 5 = 50$, then $2n = 55$, so 50 is not a term, as n must be a whole number
3 a $\frac{1}{100\,000}, \frac{1}{1\,000\,000}$ **b** $\frac{1}{100\,000\,000}$
4 a 67, 59 **b** $-8n + 107$
c -53
5 a $4, -2$ **b** 6th term
6 a $6^2 = 36$; $5^2 = 25$ **b** 10th term $= 1^2 = 1$
7 a $-2n + 14$ **b** -26
8 a 7, 10 **b** 33rd term
c $3n + 4 = 200$, $n = \frac{196}{3}$, not a term as n must be a whole number

7 Gradients of straight line graphs

7.1 Finding the gradient

1 a $\frac{2}{1} = 2$ **b** $\frac{3}{6} = \frac{1}{2}$
2 a $\frac{-10}{5} = -2$ **b** $\frac{-5 - (-3)}{6 - 4} = -\frac{2}{2} = -1$
3 a 10 **b** $3\frac{1}{3}$ **c** 5
4 a -1 **b** -3 **c** -2
5 $\frac{12}{4} = 3$
6 $-\frac{8}{4} = -2$
7 a 6 **b** 1
8 a -3 **b** -1
9 -2

7.2 Interpreting the gradient

1 a 1.2 **b** £1 = 1.20 euros
2 a £20 **b** $\frac{90}{5} = 18$
c units are cost per day, so cost per day = £18
3 a gradient = 2.2 **b** 1 kg = 2.2 pounds
4 a £2 **b** 1.8
c cost per mile = £1.80

Don't forget!

* slope
* gradient
* positive, negative, zero, infinite
* $\frac{\text{change in } y}{\text{change in } x}$
* negative
* y, x

Exam-style questions

1 1.5
2 a 30 **b** speed is 30 mph
3 -5
4 -1
5 a £20 **b** 8
c fee for each additional hour is £8
6 $\frac{5}{4}$

8 Straight line graphs

8.1 Horizontal and vertical lines

1 a $y = 3$ **b** $y = -3$ **c** $x = 1$ **d** $x = -3$
2 a $x = -1$ **b** $y = 2$ **c** $x = 2$ **d** $y = -1$
3 a $y = 0$ **b** $x = 0$
4

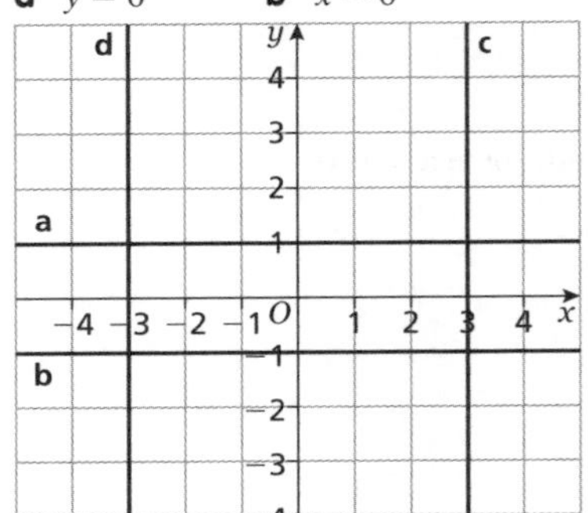

8.2 The equation $y = mx + c$

1 a gradient = 3 **b** gradient = -4 **c** gradient = 1
2 a $m = 2, c = -3$ **b** $m = -3, c = 5$ **c** $m = 7, c = 0$
3 a gradient = 8, y-intercept = 7
b gradient = 1, y-intercept = -9
c gradient = -6, y-intercept = 4
4 a $y = 4x - 3$ **b** $m = -2, c = 5$; $y = -2x + 5$
c $m = 1, c = -7$; $y = x - 7$
5 a $y = -5x + 1$ **b** $y = 7x - 2$ **c** $y = -4x$

8.3 Plotting and drawing graphs

1 a

x	-1	0	1	2	3	4
$y = 2x + 1$	-1	1	3	5	7	9

$x = -1$: $y = 2 \times -1 + 1 = -2 + 1 = -1$
$x = 0$: $y = 2 \times 0 + 1 = 0 + 1 = 1$
$x = 1$: $y = 2 \times 1 + 1 = 2 + 1 = 3$
$x = 2$: $y = 2 \times 2 + 1 = 4 + 1 = 5$
$x = 3$: $y = 2 \times 3 + 1 = 6 + 1 = 7$
$x = 4$: $y = 2 \times 4 + 1 = 8 + 1 = 9$

b

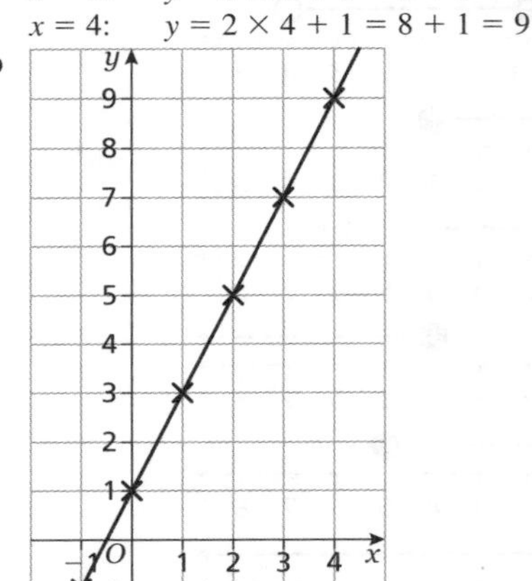

2 a

x	0	1	2	3	4	5
y	5	4	3	2	1	0

$x = 0$: $y = 5 - 0 = 5$ $x = 1$: $y = 5 - 1 = 4$
$x = 2$: $y = 5 - 2 = 3$ $x = 3$: $y = 5 - 3 = 2$
$x = 4$: $y = 5 - 4 = 1$ $x = 5$: $y = 5 - 5 = 0$

b

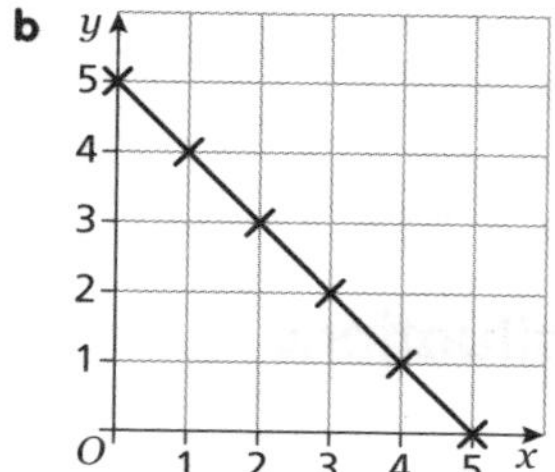

3 a

x	−1	0	1	2	3
$y = 3x - 5$	−8	−5	−2	1	4

b

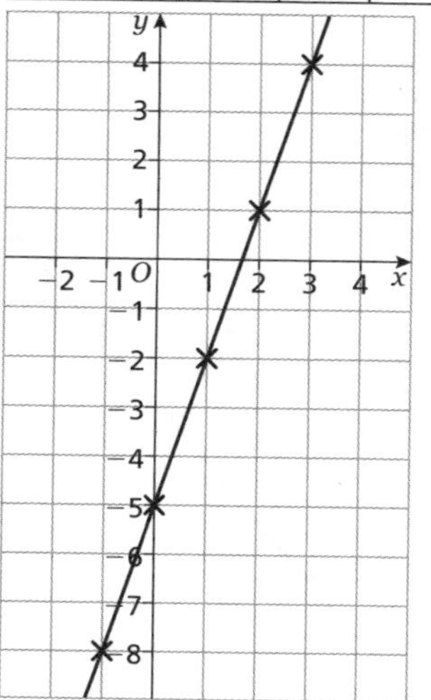

4 a

x	−1	0	1	2	3
$y = -4x + 2$	6	2	−2	−6	−10

b

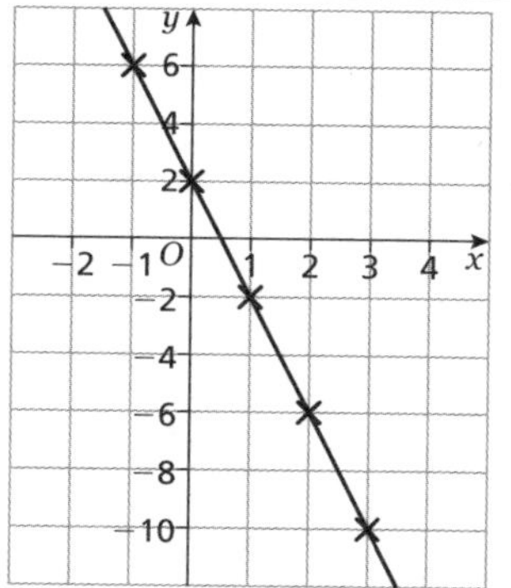

5

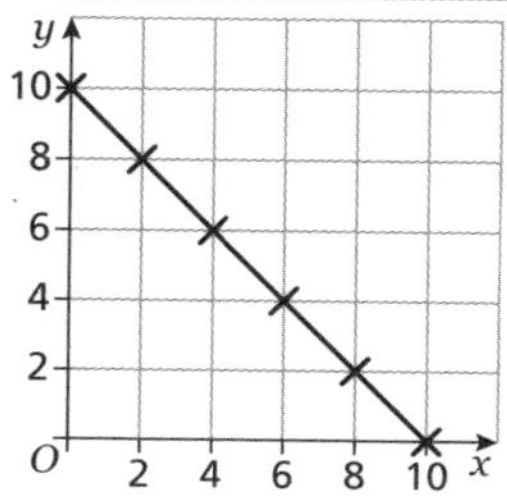

6

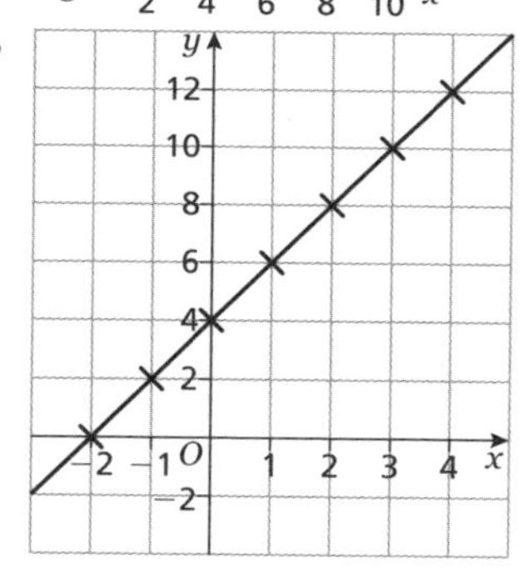

7

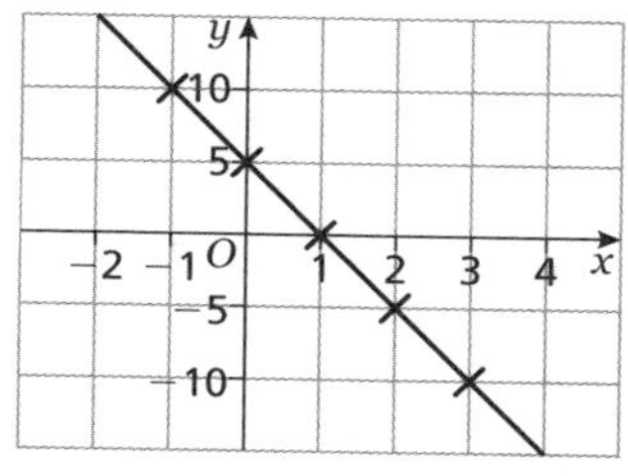

8

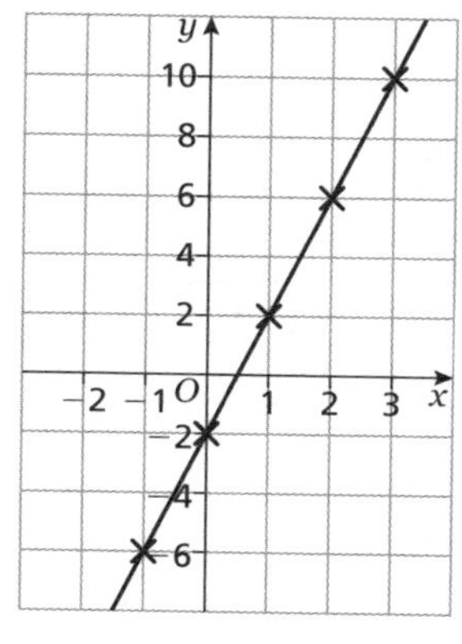

9

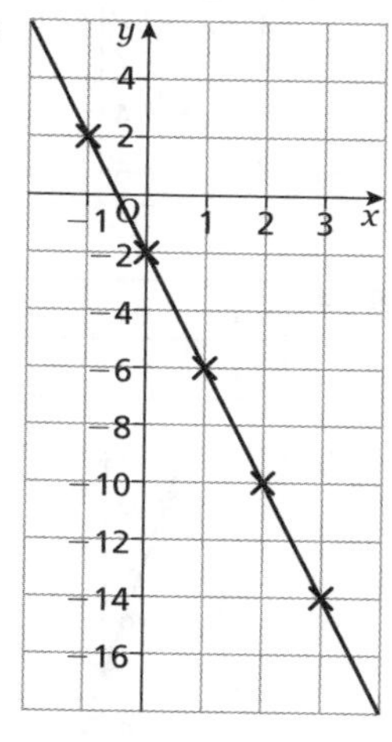

8.4 Finding the equation of a straight line graph

1 a $y = 2x - 5$

b $m = \frac{6}{2} = 3$; $y = 3x + 1$

c $c = 5, m = \frac{-2}{1} = -2$; $y = -2x + 5$

d $c = 3, m = \frac{-1}{1} = -1$; $y = -x + 3$

2 a $y = 4x - 7$
b $y = 5x$
c $y = -3x + 5$
d $y = -4x - 2$

Don't forget!

* $x = a$
* $y = b$
* $y = 0$
* $x = 0$
* m; c
* $\frac{\text{change in } y}{\text{change in } x}$
* graph A $\rightarrow y = 2x + 3$; graph B $\rightarrow y = -2x + 3$; graph C $\rightarrow y = -3x - 1$; graph D $\rightarrow y = 4x + 1$

Exam-style questions

1

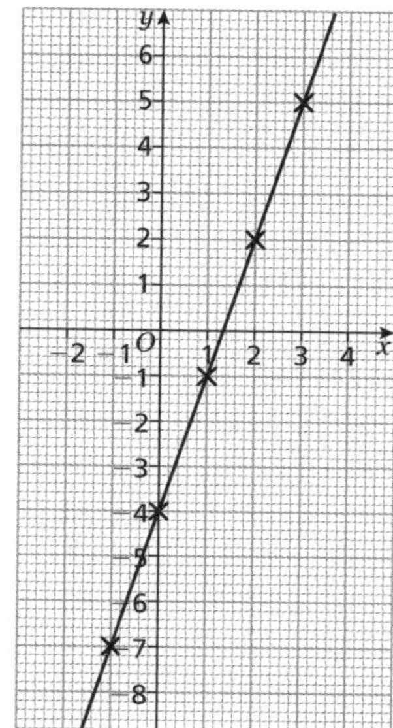

2 a 4
b $y = 4x - 1$

3

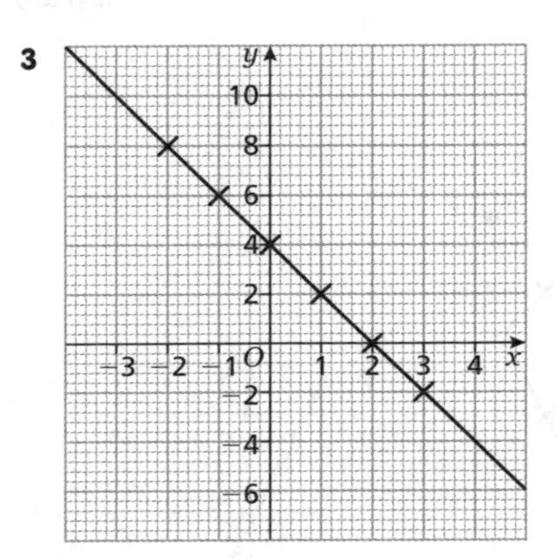

4 a

x	-1	0	1	2	3	4
y	10	7	4	1	-2	-5

b

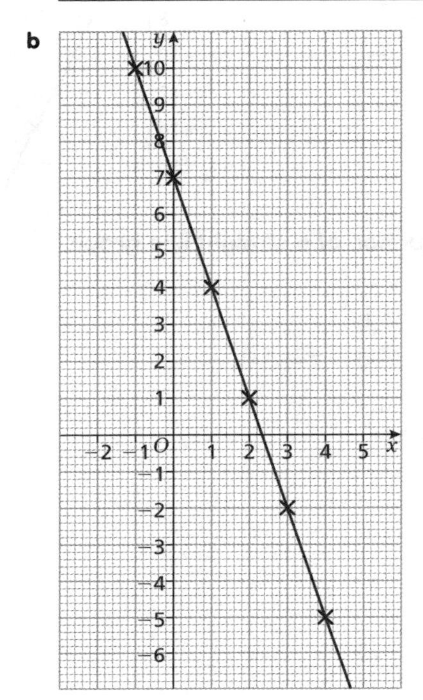

5

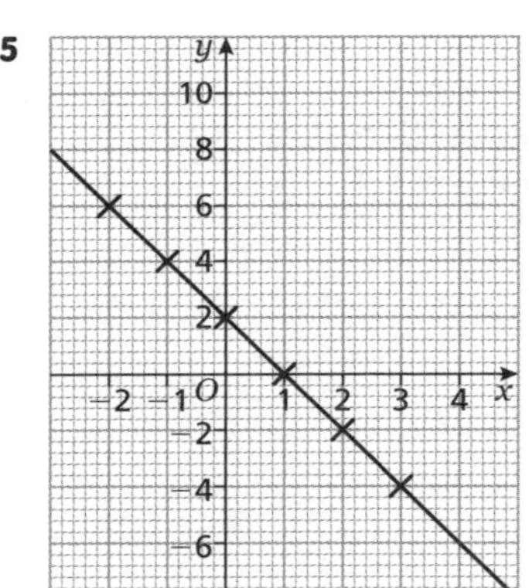

6

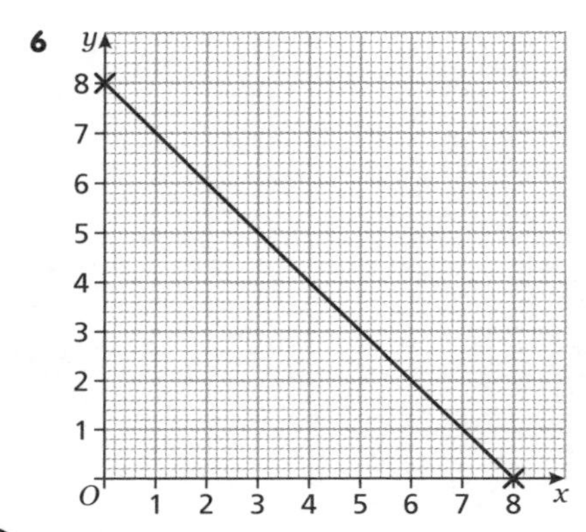

7

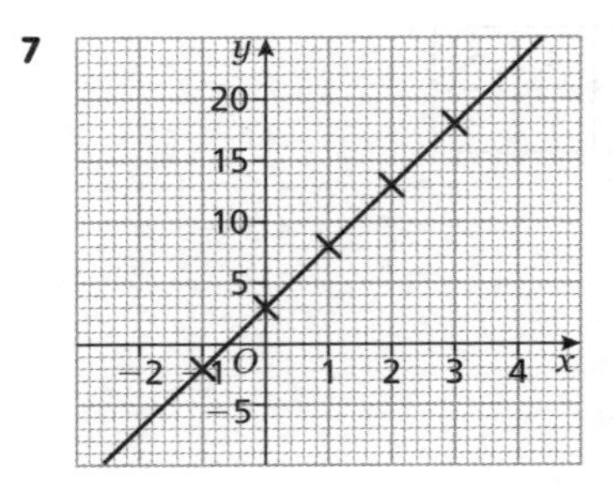

9 Graphs for real-life situations

9.1 Straight line graphs

1 a

Pints	0	10	20	30	40	50
Litres	0	6	12	18	24	30

1 pt = 0.6 litre
10 pt = 10×0.6 = 6 litres
20 pt = 20×0.6 = 12 litres
30 pt = 30×0.6 = 18 litres
40 pt = 40×0.6 = 24 litres
50 pt = 50×0.6 = 30 litres

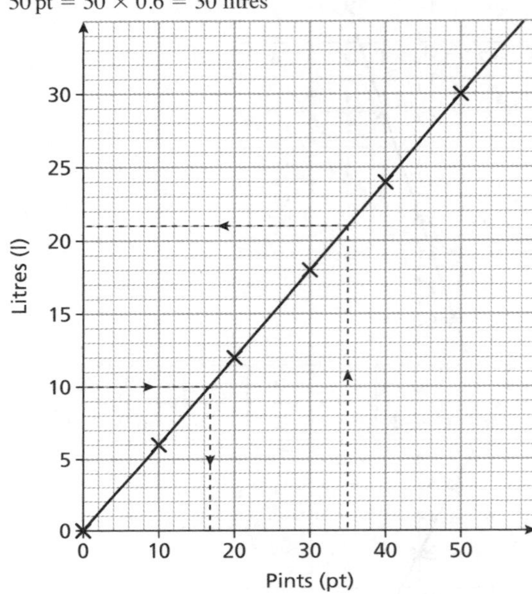

b 21 litres 
c 17 pints

2 a

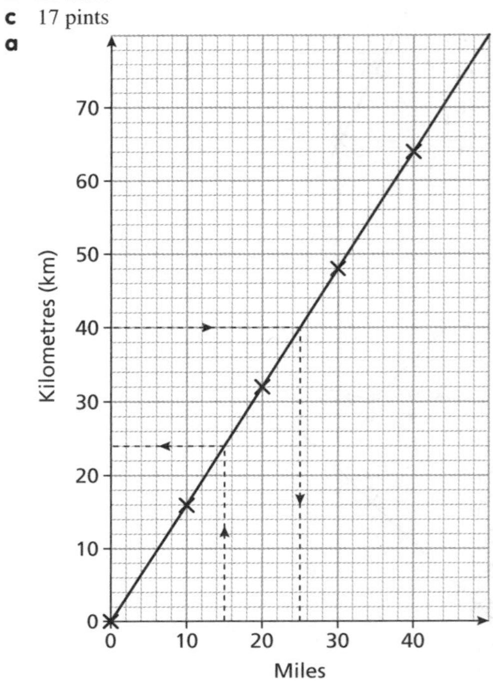

b 24 km
c 25 miles

3 a

Journey distance (miles)	0	5	10	15	20	25
Cost in pounds (£)	4	14	24	34	44	54

0 miles: cost = £4
5 miles: cost = $4 + 5 \times 2$ = £14
10 miles: cost = $4 + 10 \times 2$ = £24
15 miles: cost = $4 + 15 \times 2$ = £34
20 miles: cost = $4 + 20 \times 2$ = £44
25 miles: cost = $4 + 25 \times 2$ = £54

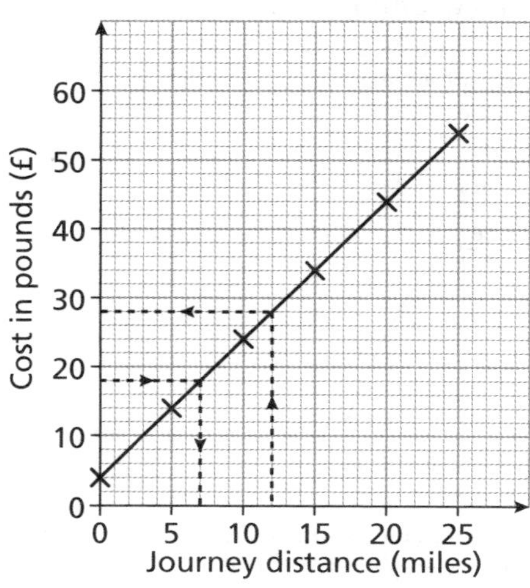

b £28
c 7 miles

4 a

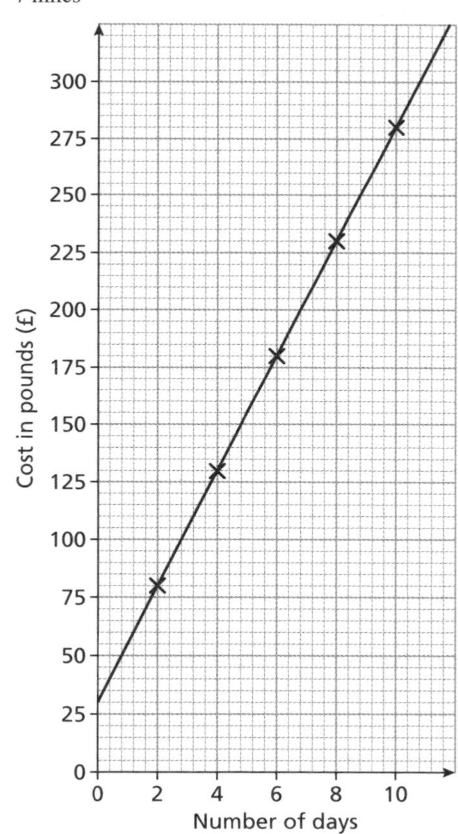

b £205
c 9 days

5 a £50
b $\frac{50}{2}$ = £25/hour
c 2
d electrician B charges £150, so Kerry should choose electrician B

6 a

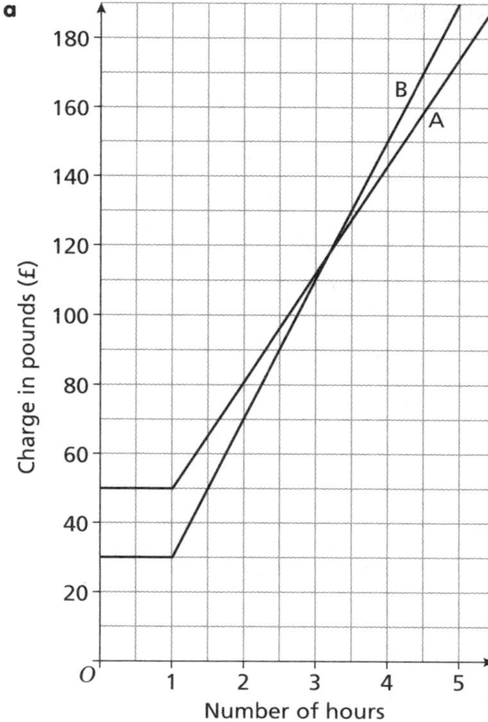

b 3 hours
c Plumber A is cheaper; A costs £140, B costs £150

7 a From top to bottom: B, C, A
b Gym C, £44
c Up to 4 sessions gym C is the cheapest. For 5 or more sessions gym A is the cheapest at £45

9.2 Different graph shapes

1 Graph C
2 1 C, 2 A, 3 B
3

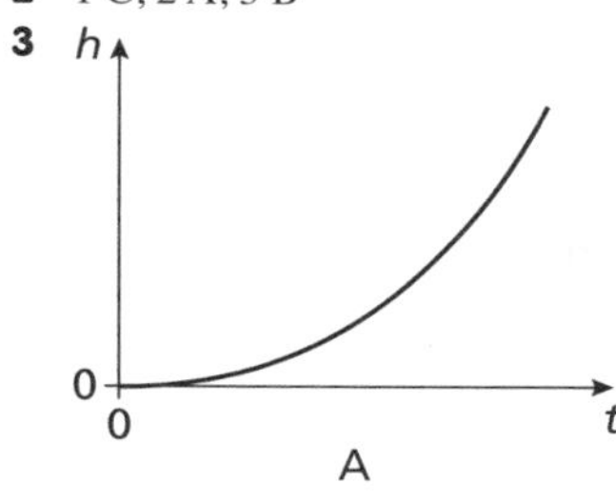

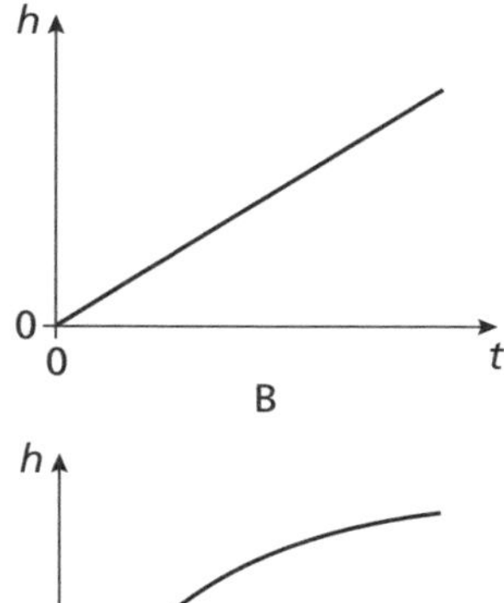

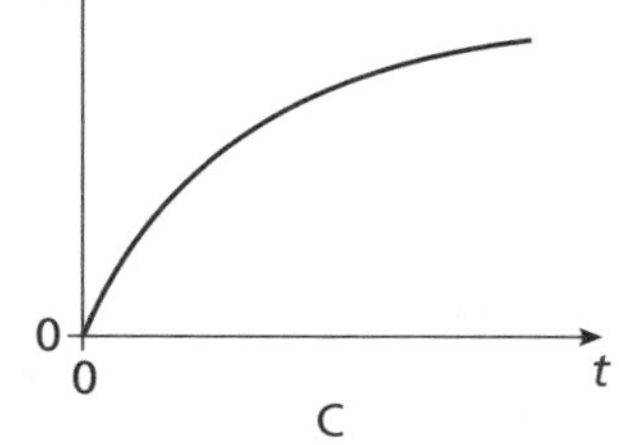

Don't forget!

* gradient; axes * increases * decreases
* Graph A → positive gradient; as one unit increases, the other increases
Graph B → negative gradient; as one unit increases, the other decreases

Exam-style questions

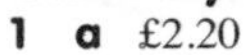

1 **a** £2.20 **b** 250 grams

2 **a**

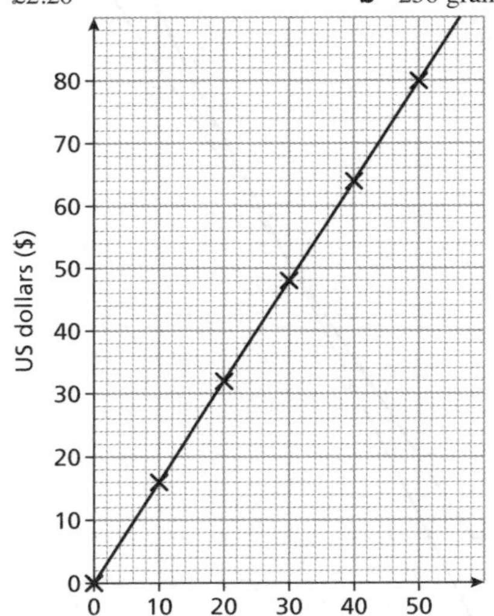

b $56 **c** £40

3 Graph D (the coffee will cool down to room temperature)

4 **a** 30 cm **b** 6 hours **c** 22.5 cm

5 **a** From top to bottom: C, A, B

b Company A, £100

c Company C would be cheapest at £155; A and B are both £160

10 Graph sketching

10.1 Quadratic graphs

1 **a** Not a quadratic

b Quadratic and ∪ shape as coefficient of x^2 is 2

c Quadratic and ∩ shape as coefficient of x^2 is -1

d Not a quadratic as highest power is x^3

2 **b** and **d**

3 **a** ∩ **c** ∪

4 **a** and **c**

10.2 Graphs of the form $y = ax^2$

1 **a**

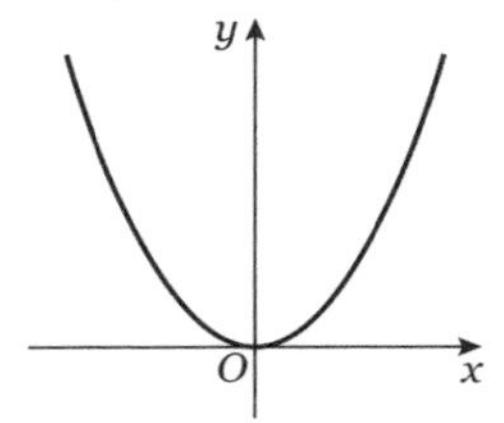

b Minimum value of $y = 0$

2 **a**

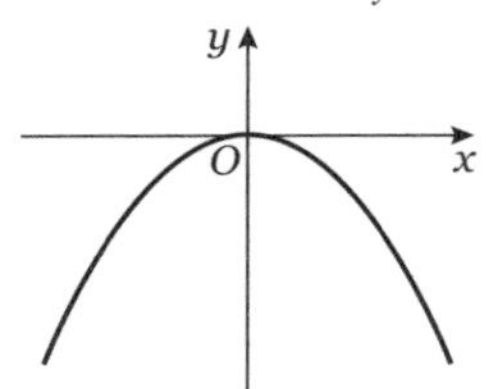

b Maximum value of $y = 0$

3 Both graphs are ∪ shaped as the coefficient of x^2 in both equations is +.
Both graphs pass through the point (0, 0).
To find the y values for the graph of $y = 3x^2$, multiply each of the values of y from the graph of $y = x^2$ by 3.

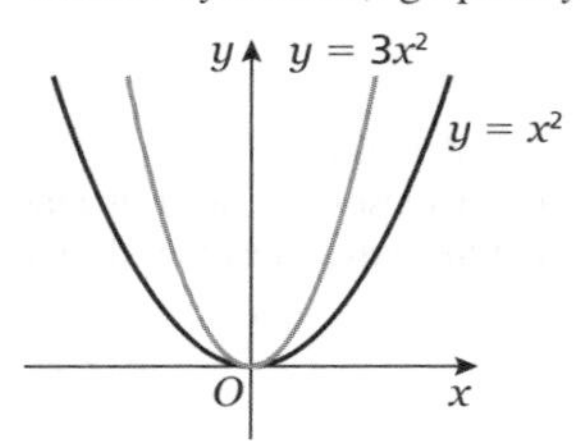

4

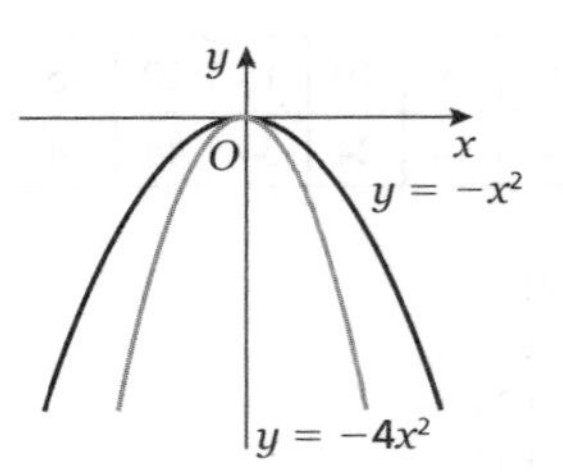

10.3 Graphs of the form $y = ax^2 + b$

1 **a** The graph crosses the y-axis when $x = 0$, so $y = 0 - 9 = -9$
The graph passes through the point $(0, -9)$.

b Minimum value of $y = -9$

c $x = 3$ and -3

2 **a**

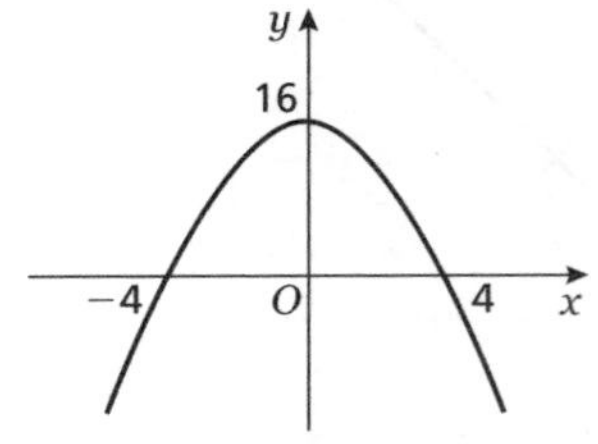

b Maximum value of $y = 16$

c $(-4, 0)$ and $(4, 0)$

3 **a** To find the y values for the graph of $y = -2x^2$, multiply the y values from the graph of $y = -x^2$ by 2.
The graph of $y = -2x^2 - 3$ is a translation of the graph of $y = -2x^2$ by -3 units in the y-axis.
The graph will pass through the point $(0, -3)$

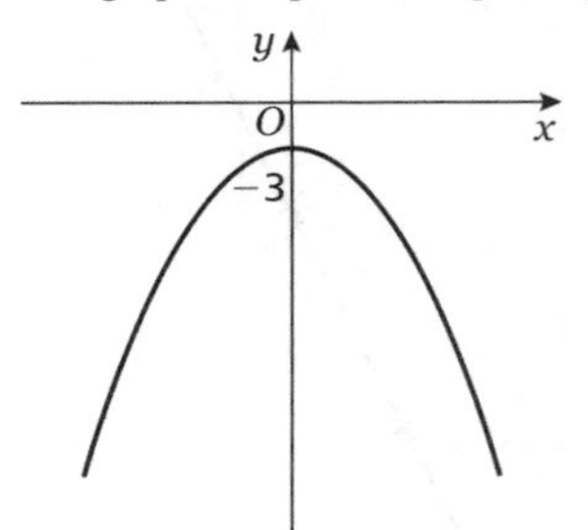

b Maximum value of $y = -3$

4

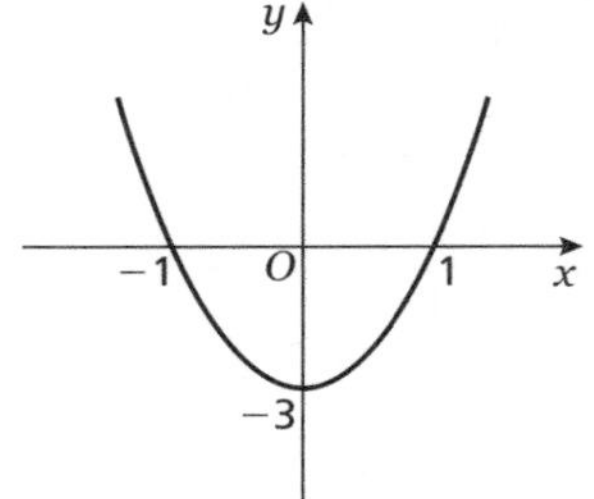

10.4 Graphs of the form $y = (x + b)^2$

1 **a** $(0, 4)$

b $(-2, 0)$

c

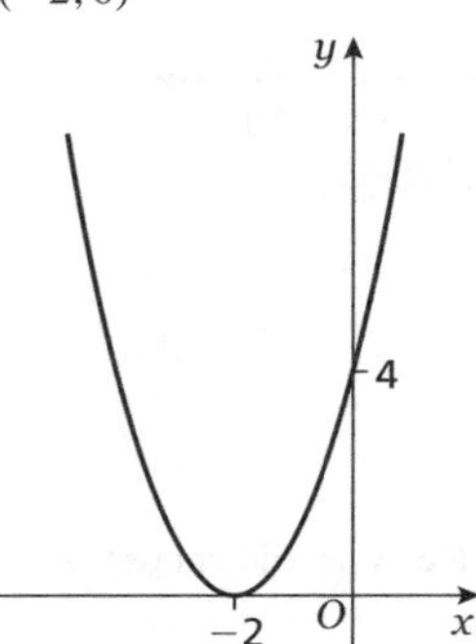

d $x = -2$

e $(-2, 0)$

2 a $y = 9$, so curve passes through $(0, 9)$
b $x = 3$, so curve meets the x-axis at $(3, 0)$
c

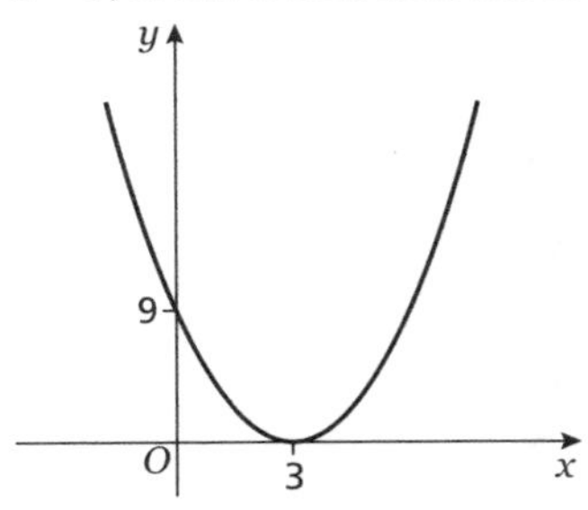

d $x = 3$
e $(3, 0)$

3 a $(0, 16)$
b $(-4, 0)$
c

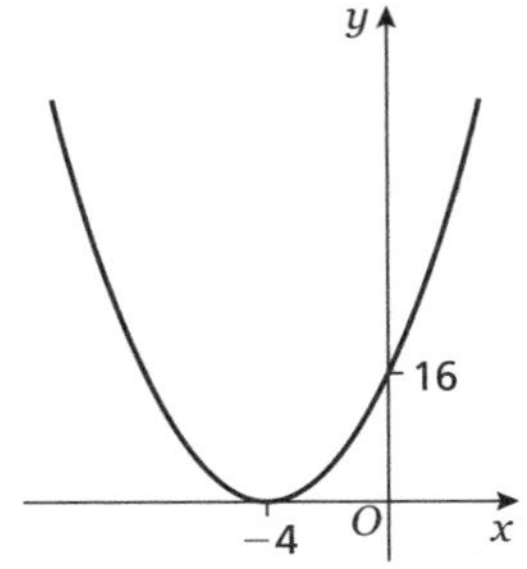

d $x = -4$
e $(-4, 0)$

4 a $(0, 36)$
b $(6, 0)$
c

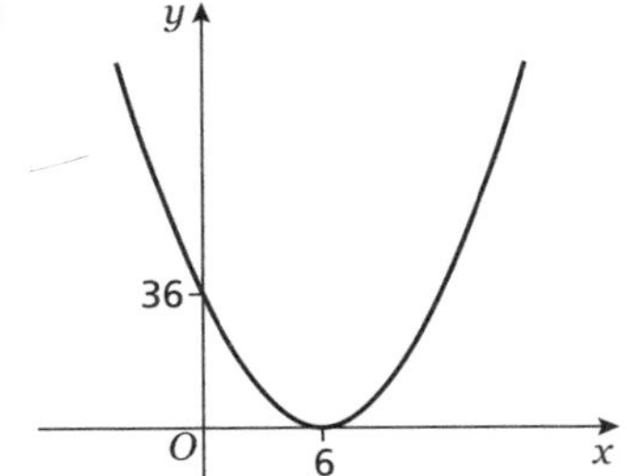

d $x = 6$
e $(6, 0)$

Don't forget!

* x^2
* parabola
* x^2
* $+$
* $-$
* symmetry
* $x = 0$
* $y = 0$
* $(0, 0)$; 0; multiplying
* $(0, 5)$; 5; 5, y
* b
* $(0, 9)$; $(-3, 0)$; $x = -3$; 3, left

Exam-style questions

1 a

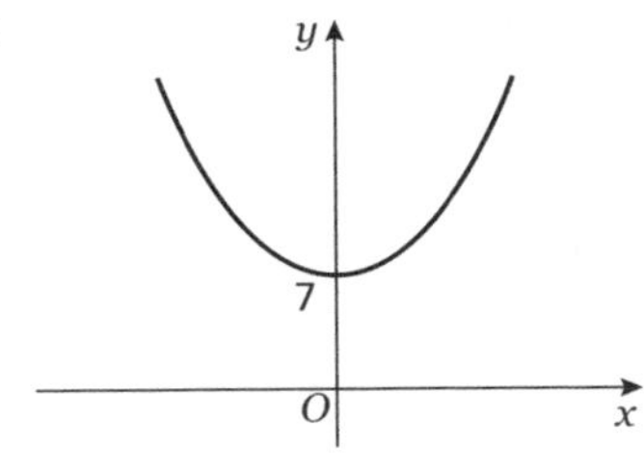

b 7

2

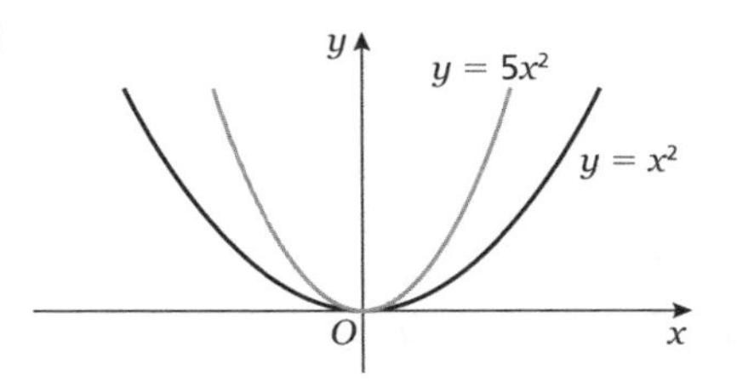

3 a

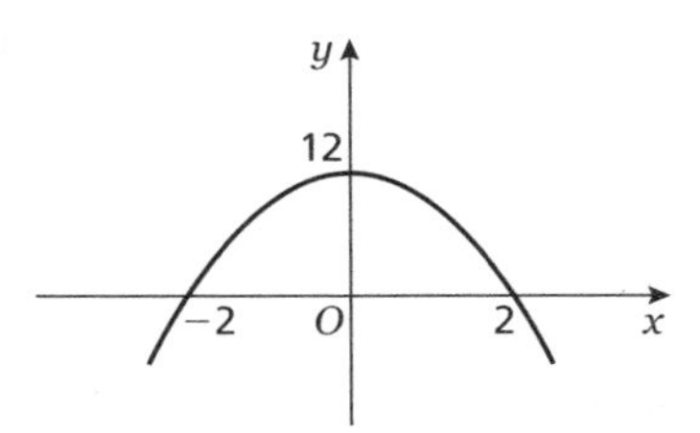

b 12

4 a

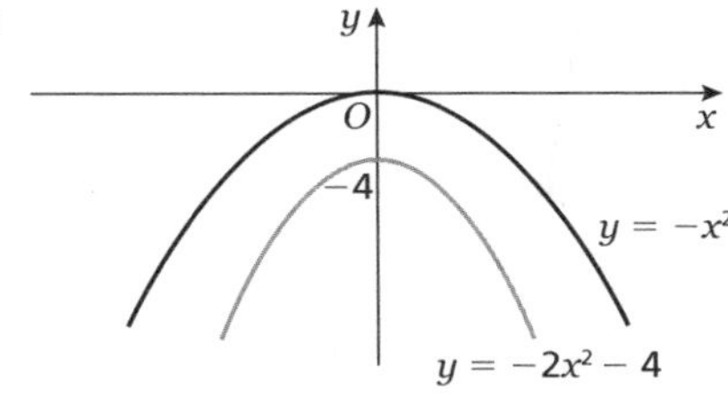

b 0
c -4

5 a $(0, 25)$
b $(5, 0)$
c

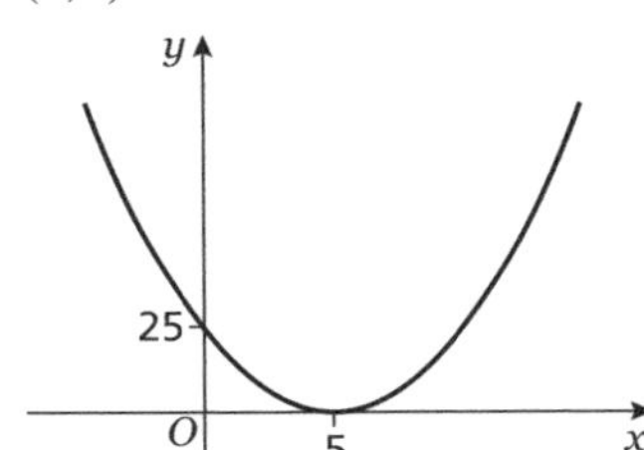

d $x = 5$

11 Simple quadratic functions

11.1 Plot graphs of quadratic functions of the form $y = ax^2 + b$

1 a

x	-3	-2	-1	0	1	2	3
$y = x^2$	9	4	1	0	1	4	9

$x = -1$: $y = (-1)^2 = 1$
$x = 0$: $y = (0)^2 = 0$
$x = 1$: $y = (1)^2 = 1$
$x = 2$: $y = (2)^2 = 4$
$x = 3$: $y = (3)^2 = 9$

b

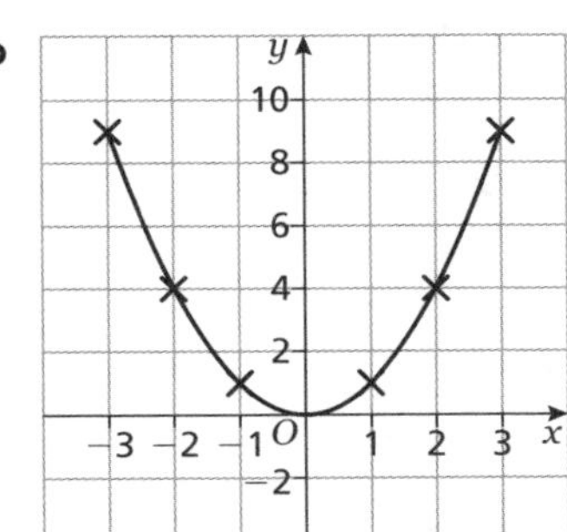

2

x	−3	−2	−1	0	1	2	3
$y = -x^2$	−9	−4	−1	0	−1	−4	−9

x	−3	−2	−1	0	1	2	3
$y = -2x^2$	−18	−8	−2	0	−2	−8	−18

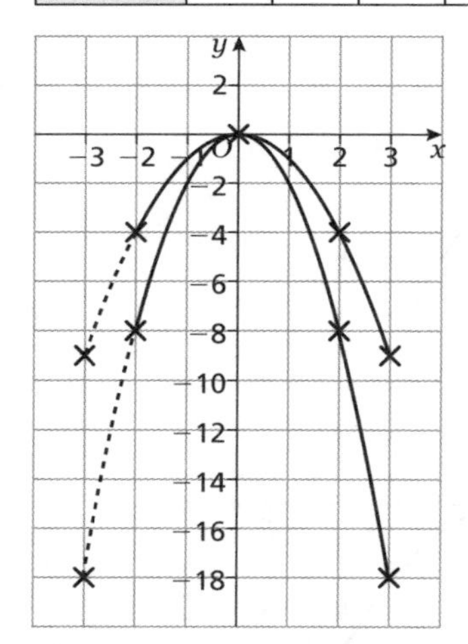

3

x	−3	−2	−1	0	1	2	3
$y = x^2 + 2$	11	6	3	2	3	6	11

$x = -1$: $y = (-1)^2 + 2 = 3$
$x = 0$: $y = (0)^2 + 2 = 2$
$x = 1$: $y = (1)^2 + 2 = 3$
$x = 2$: $y = (2)^2 + 2 = 6$
$x = 3$: $y = (3)^2 + 2 = 11$

x	−3	−2	−1	0	1	2	3
$y = x^2 - 3$	6	1	−2	−3	−2	1	6

$x = -1$: $y = (-1)^2 - 3 = -2$
$x = 0$: $y = (0)^2 - 3 = -3$
$x = 1$: $y = (1)^2 - 3 = -2$
$x = 2$: $y = (2)^2 - 3 = 1$
$x = 3$: $y = (3)^2 - 3 = 6$

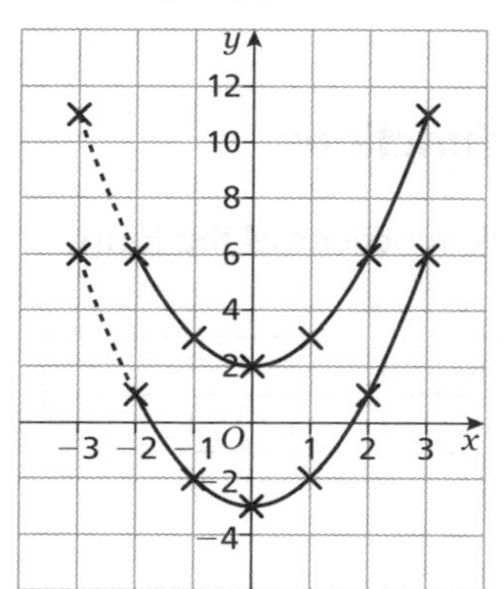

4 a

x	−3	−2	−1	0	1	2	3
$y = 3x^2$	27	12	3	0	3	12	27

b

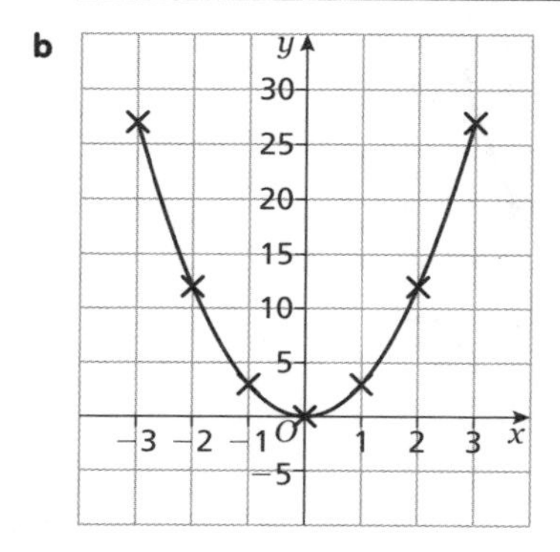

5

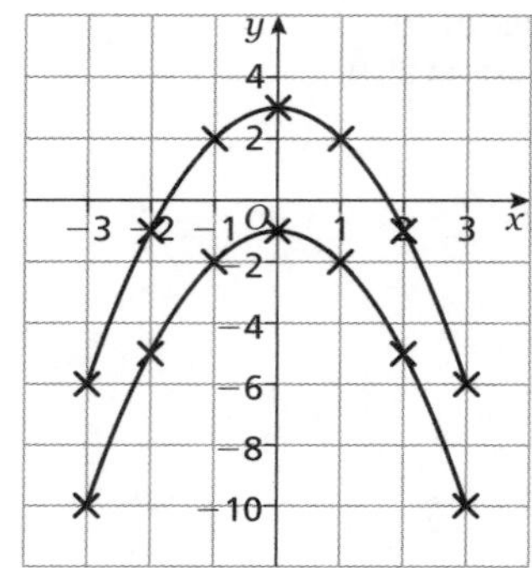

6 a

x	−3	−2	−1	0	1	2	3
$y = 4x^2 - 5$	31	11	−1	−5	−1	11	31

b

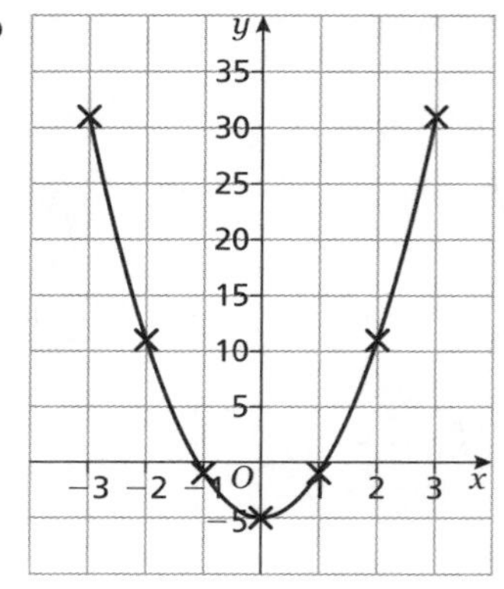

11.2 Plot graphs of quadratic functions of the form $y = ax^2 + bx + c$

1 a

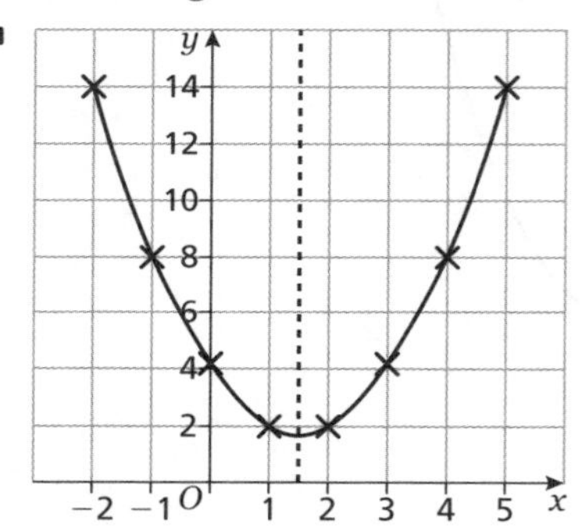

x	−2	−1	0	1	2	3	4	5
x^2	4	1	0	1	4	9	16	25
$-3x$	+6	+3	0	−3	−6	−9	−12	−15
$+4$	+4	+4	+4	+4	+4	+4	+4	+4
$y = x^2 - 3x + 4$	14	8	4	2	2	4	8	14

b $x = 1.5$

2 a

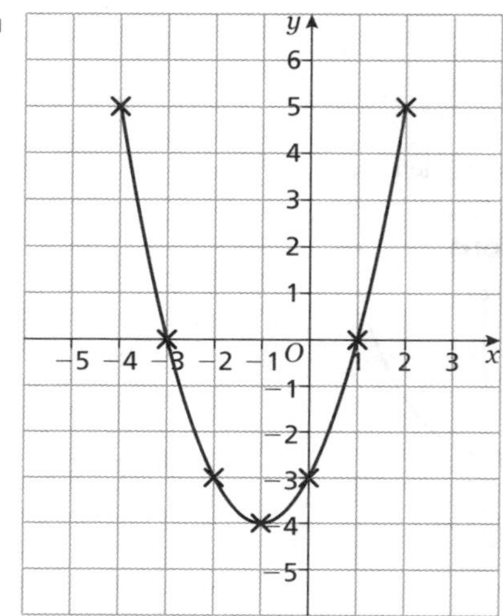

b $x = -1$

3 a

x	−2	−1	0	1	2	3	4
y	8	4	2	2	4	8	14

b

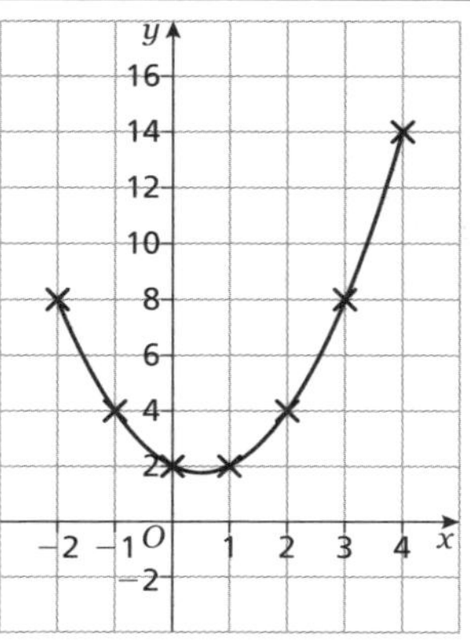

c $x = 0.5$

4 a

x	0	1	2	3	4	5
y	0	−3	−4	−3	0	5

b

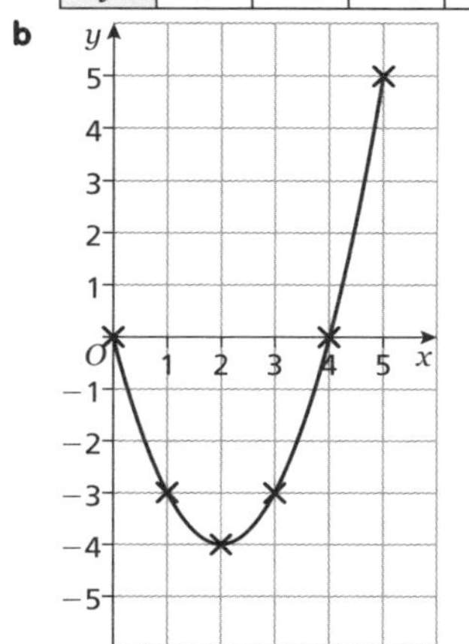

11.3 Using quadratic graphs to solve equations

1 a $x = -0.3$ and $x = 3.3$
b $x = -1.5$ and $x = 4.5$
c $x = -0.8$ and $x = 3.8$
2 a $x = -2.8$ and $x = 1.8$
b Draw $y = -2$: $x = -2.3$ and $x = 1.3$
c Draw $y = 4$: $x = -3.5$ and $x = 2.5$
3 a $x = -4.4$ and $x = 0.4$
b Draw $y = -4$: $x = -3.4$ and $x = -0.6$
c Draw $y = 2$: $x = -4.8$ and $x = 0.8$

Don't forget!

* curve
* positive
* x-axis
* $x^2 - 5x$; 4
* line, symmetry
* $(0, c)$
* $y = k$; cross

Exam-style questions

1

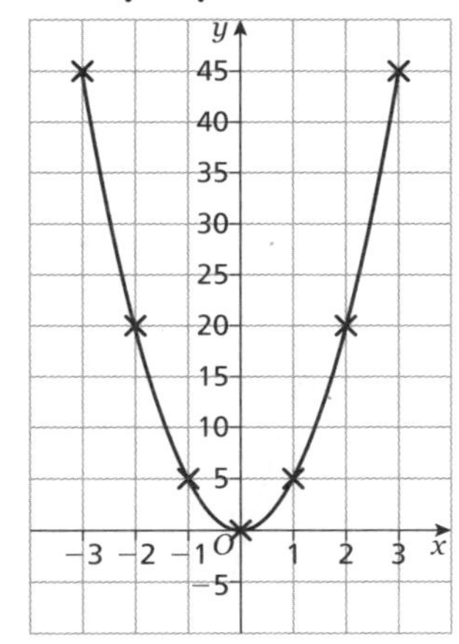

2

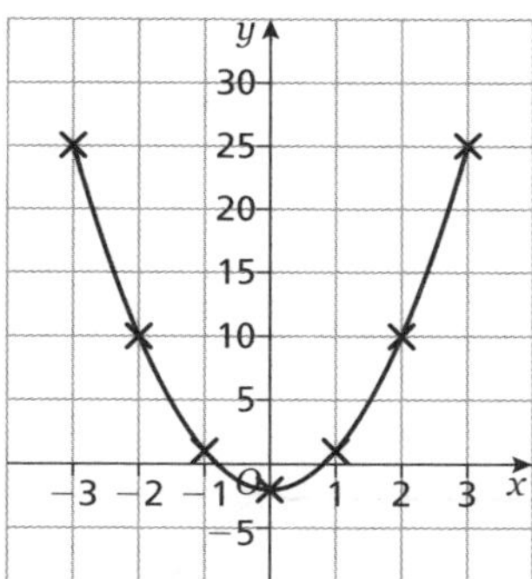

3 a

x	−3	−2	−1	0	1	2	3
y	−15	−5	1	3	1	−5	−15

b

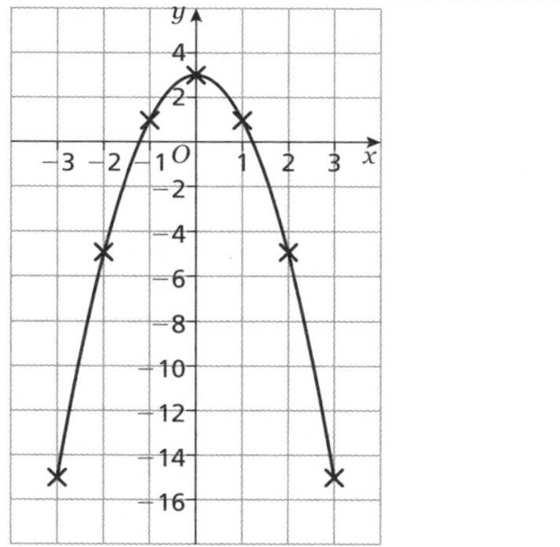

4 a

x	−2	−1	0	1	2	3	4
y	4	−1	−4	−5	−4	−1	4

b

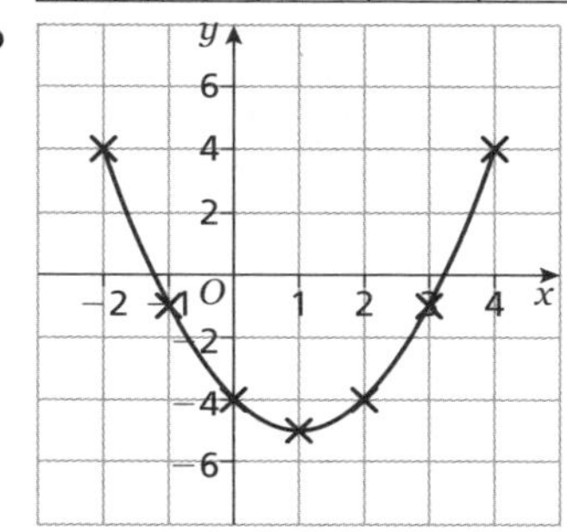

c $x = 1$
d $x = -1.2$ and $x = 3.2$
e Draw $y = 3$, $x = -1.8$ and $x = 3.8$

5 a

x	0	1	2	3	4	5	6
y	0	−5	−8	−9	−8	−5	0

b

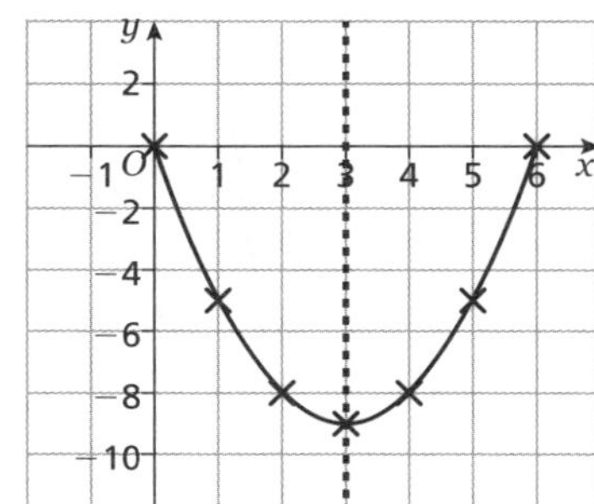

c $x = 3$
d Draw $y = -6$: $x = 1.3$ and $x = 4.7$

12 Distance–time and speed–time graphs

12.1 Speed

1 40 mph
2 4 km/h
3 130 km/h
4 80 mph
5 30 km/h

12.2 Distance–time graphs

1 **a** $60 \div 4 = 15$ min; Kai leaves home at 15:15
b 3 km
c 15:45
d 60 min; 6 km/h
e 17:45

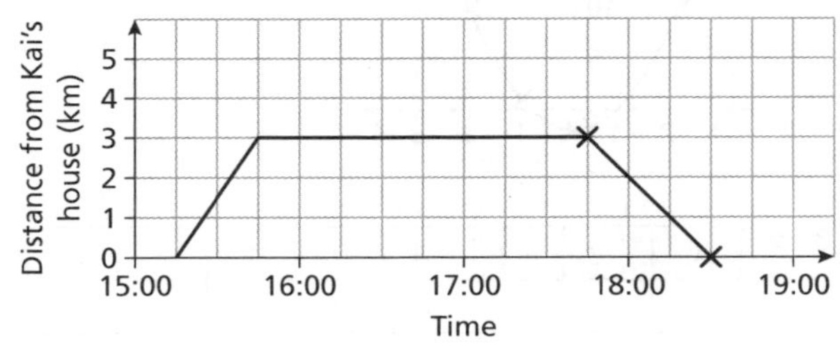

f to; to; Check: 30 min, 45 min.

2 **a** 12:40
b 16 km
c 32 km/h
d 90 min or 1.5 hours
e 24 km/h

3 **a** Becky travels 50 miles in 2 hours, then stops for 30 minutes. She then travels 40 miles in 1 hour to see her friend and stops for 3.5 hours. Becky then takes 2 hours to travel 90 miles back to her starting point.

b

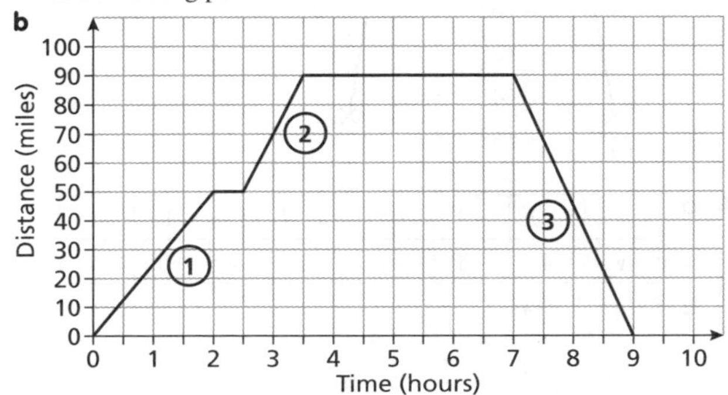

(1) 25 mph
(2) 40 mph
(3) 45 mph

4 **a** (35, 7.5)

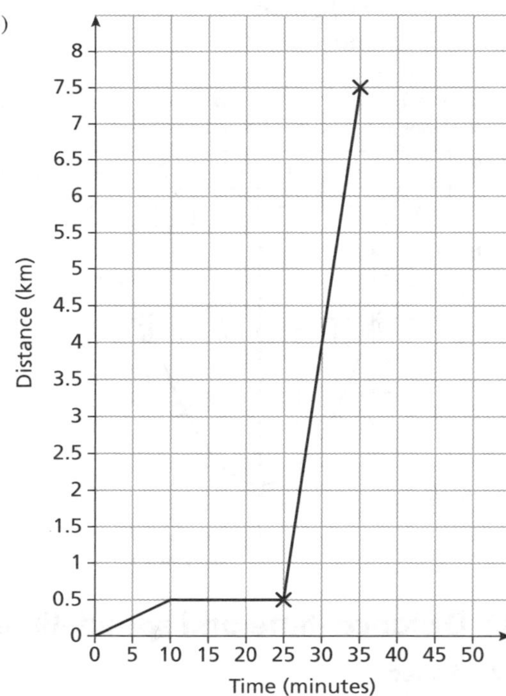

b 42 km in 60 min; Average speed = 42 km/h

5 **a**

Distance from Joe's home (km) — 0, 5, 10, 15, 20; Time — 2 pm, 3 pm, 4 pm, 5 pm

b 40 km/h

12.3 Speed–time graphs

1 **a** 2 minutes = **120** seconds

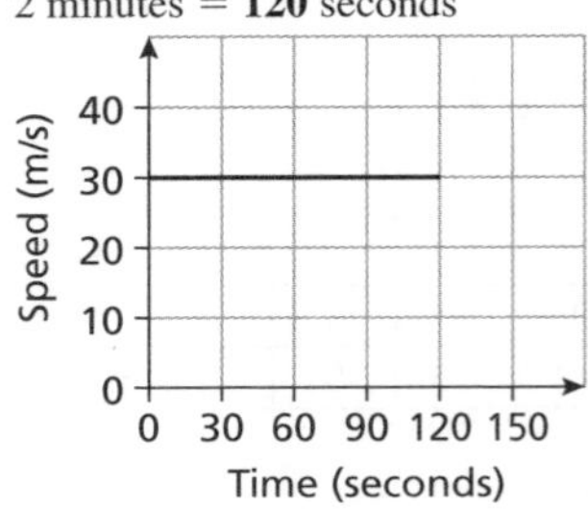

b speed × time = distance
distance = speed × time
distance = $30 \times 120 = 3600$ m

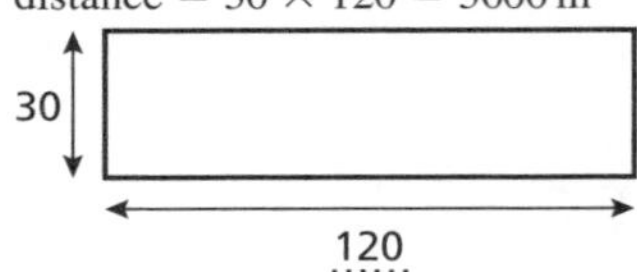

2 **a** (3) The scooter goes from 12 m/s to 0 in 5 seconds
b Area of triangle **A** = 30 m;
Area of rectangle **B** = $10 \times 12 = 120$ m;
Area of triangle **C** = 30 m;
Total distance between traffic lights = $30 + 120 + 30 = 180$ m

3 **a**

Speed (m/s) — 0, 1, 2, 3, 4, 5, 6, 7, 8; Time (seconds) — 0, 2, 4, 6, 8, 10, 12, 14

b 64 m

4 **a** Car goes from 0 to 20 m/s in 10 seconds, then travels at a constant speed of 20 m/s for 15 seconds. Car then goes from 20 m/s to 0 in 15 seconds
b 550 m

Don't forget!

* Average speed
* constant speed; horizontal; the speed; faster
* horizontal; positive; negative; distance

Exam-style questions

1 **a** Barry stops for 20 minutes
b

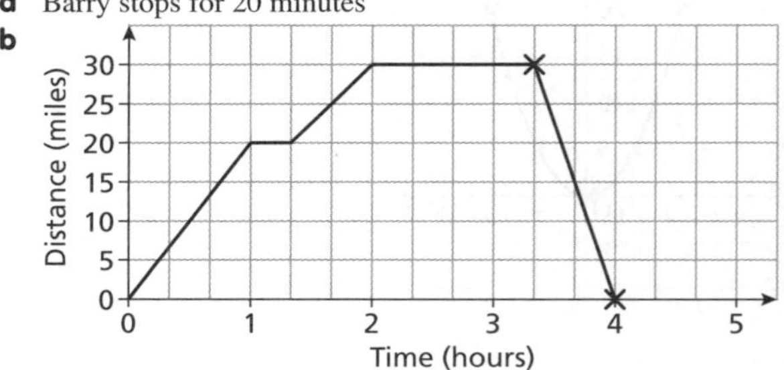

2 a

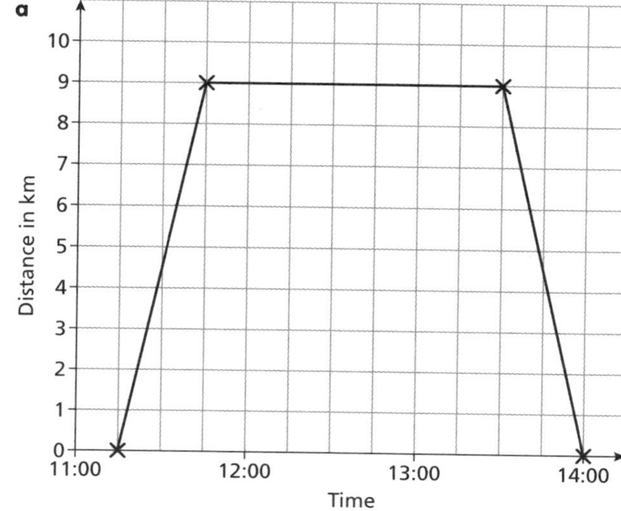

b 18 km/h

3 a

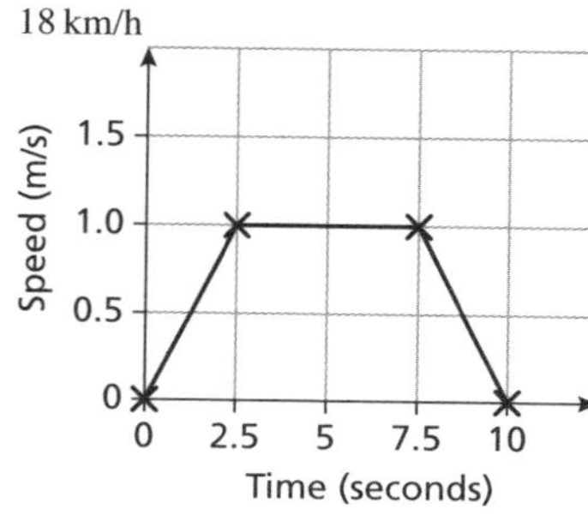

b 7.5 m

4 a The truck goes from 0 to 10 m/s in 5 seconds, then travels at a constant speed of 10 m/s for 7 seconds. The truck then goes from 10 m/s to 0 in 2 seconds.

b 105 m

Practice Paper

1 $3s + 5l$

2 $V = IR$; $C = 2\pi r$; $A = \frac{1}{2}bh$

3 a b^8 **b** $4p^4$ **c** $8k^6$ **d** $12x^6y^4$

4 a £125 **b** 40

5 a $3g^2 - 8h^2 + 8$ **b** $24s^2 - 28$

c $6w^3 - 15w^2 - 3w$ **d** $-2v^2 + 13v$

6 a 1 **b** 6 **c** $\frac{y - c}{m}$

7 a 37

b No; explanation showing non-integer solution, e.g.:

$4n - 3 = 200$

$4n = 200 + 3$

$n = 203 \div 4 = 50.75$, which is not an integer

8 a $4x(y + 2)$ **b** $2a^2b(a^2 - 2ab + 3)$

9 a $b = 23$ **b** $w = \frac{1}{4}$ **c** $v = -\frac{1}{2}$

10

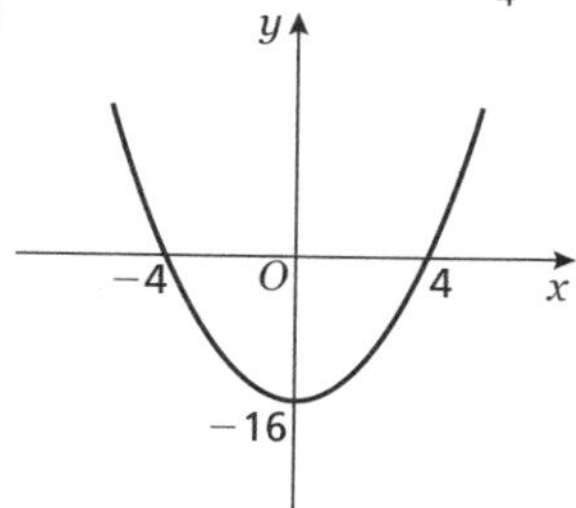

11 a $-3, -2, -1, 0, 1, 2$

b 5

12 a -2

b $y = -2x + 3$

13 a Lines labelled from top to bottom: B, A, C

b Gym B is the cheapest up to 3 months
(Gym A cost = $20 + 15 \times 3$ = £65;
Gym B cost = 20×3 = £60;
Gym C cost = $35 + 12 \times 3$ = £71;
Gyms A and B cost the same at 4 months (£80, compared with £83 for Gym C);
Gyms A and C cost the same at 5 months (£95, compared with £100 for Gym B);
Gym C is the cheapest at 6 months and longer (£107 for 6 months, compared with £110 for Gym A and £120 for Gym B)

14 a Train travels at 60 m/s for first 15 seconds, then slows down to a stop between 15 and 20 seconds.

b 1050 m

15 a

x	-2	-1	0	1	2	3	4	5
y	5	-1	-5	-7	-7	-5	-1	5

b

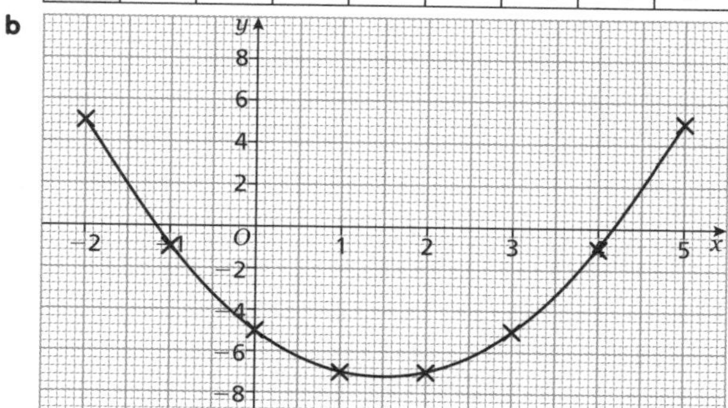

c $x = 4.2 \pm 0.2$, $x = -1.2 \pm 0.2$

d Draw $y = -3$: $x = 3.6 \pm 0.2$, $x = -0.6 \pm 0.2$